Amazing Space Facts For Curious Kids

1000 Facts Happening Above Your Head Right Now that Will Blow Your Mind

Matt J. Kessler

Table of Contents

Introduction

Have you ever been curious about what lies in the universe beyond our planet? When did everything start in the world?

There are a lot of unanswered questions about the cosmos. Therefore, it should not come as a surprise to any of us that we are all so fascinated by it and are always trying to learn more about it. We are curious about the world we reside in, aren't we?

There is a vast amount of information about the cosmos that we have just started to scratch the surface of learning. However, most details about how the universe operates are still unknown to scientists. Because of the rapid advancement of our technology, we can now expand our knowledge of the cosmos and formulate hypotheses to explain some of the most perplexing issues ever posed.

As a result, this book is written to give you information about the universe's entirety! This list of one thousand incredible facts about space and the universe will appeal to you if you are interested in things such as planets, rockets, chocolate, and corny jokes. So, fasten your seatbelts and take pleasure in the voyage to Planet Fact!

As a kid, we are constantly fascinated by the vastness of the universe. Equally, many individuals are mesmerized by the elements of the universe, and they have, for an exceptionally long time, possessed a natural curiosity regarding its various secrets. The term "universe" refers to the totality of all things that can be found in space. It is home to countless galaxies, stars, gas clouds, and even black holes, amongst other wondrous things. These are some of the most amazing facts about the universe, and you will find them fascinating because there is still so much about space, the galaxy, and our solar system, that we don't know!

Space is huge. Because billions of galaxies, stars, and planets in our solar system have not been thoroughly investigated or comprehended, the scientific community's understanding of space is always expanding. However, there are some extremely cool things that scientists have discovered about space recently. Consequently, I have produced a list of what I believe to be one thousand astounding facts, which I hope you will consider to be out of this world. Happy reading!

Chapter 1: Facts About the Universe

"Space is for everybody. It's not just for a few people in science or math or for a select group of astronauts. That's our new frontier out there, and it's everybody's business to know about space." – **Christa McAuliffe**

The Number of Stars in Space is Unknown

It is impossible to estimate the precise number of stars in our universe. However, scientists and astronomers believe that the Milky Way alone has 200 to 400 billion stars at this time. As a result, our cosmos contains billions of galaxies.

Astronomers estimate that the Milky Way alone contains about 100,000,000,000 stars. Outside of that, there are billions upon billions of other galaxies! A comparison has been made between trying to count the stars in the Universe and depending on the sand grains on an Earthly beach. To do that, we could determine the beach's size and the sand layer's typical depth. We can calculate the total number of grains on the beach by multiplying the number of grains in a small, representative sample of sand.

The galaxies act as our scale models of the Universe; our galaxy contains about 1011–1012 stars, and there are 1011–1012 galaxies overall. According to this simple calculation, the Universe has an estimated 1022–1024 stars. It is an approximation because, as is evident, no two galaxies are alike, just as different parts of a beach won't have the same sand depth. We measure integrated quantities such as the number and luminosity of galaxies rather than counting stars one by one.

ESA's infrared satellite observatory Herschel has made a significant contribution by making the first-ever attempt at "counting" galaxies in the infrared and calculating their luminosity in this range. Knowing the rate of star formation makes calculations more precise. Herschel has also plotted the speed at which stars have formed throughout cosmic history. By estimating the rate at which stars are produced, you can calculate the number of stars in the Universe.

The Hubble Space Telescope (HST) captured an image in 1995, showing that star formation peaked around 7,000,000,000 years ago. Astronomers have since had a change of heart. We now have evidence that thick dust clouds hid most early star formations. The stars are hidden by dust clouds, which convert their light into infrared radiation, preventing the HST from seeing them. At infrared wavelengths, Herschel was able to observe this previously unobserved Universe and discovered a vast number of previously unknown stars.

There is Total Silence in Space

Since there is no atmosphere in space, sound cannot travel, so no one hears a sound. However, since radio waves can still be sent and received in space, astronauts utilize radios to stay in touch. Anyone who paid attention in class would tell you that vacuums do not conduct sound and that space is one such vacuum. The first portion of this statement raises some doubts, even though the second part is unquestionably accurate.

After all, a vacuum is an object that contains no matter at all. It has absolutely nothing, not even a single particle. However, given that there are planets, stars, clouds of dust, and substances in space, how can this assertion about space be true? But hold off on blaming your high school physics course. According to what we now know about space, most of it is a vacuum. Space is already too big a place. All celestial bodies are separated from one another by vacuum. In other words, albeit not entirely, space is a vacuum.

The Chinese Started Recording the Sighting of Halley's Comet in 240B

It was continuously recorded every time the comet was visible, starting in 164 BC. In 2061, the next Halley's comet will be visible from Earth. Chinese astronomers spot a brand-new "star" in the sky that resembles a broom around 240 B.C. It is Halley's Comet's first verified sighting.

Some have argued that the alignment of the Giza pyramids and the Sphinx is due to a sighting in the third century B.C. Interesting. Even a Chinese observation reported in 613 B.C. would have occurred seven years after the estimated 620 B.C. for Halley's passage. Was that a different comet or a record-keeping error?

The 240 B.C. observation agrees with Halley's calculated orbit, but its precise date is uncertain. More than a century afterward, The Records of the Grand Historian, also known as Shiji (or Shi Chi), approximately 100 B.C., is the only Chinese document still in existence. Due to its bristly tail, this object earned the nickname "broom star" among the Chinese. It first emerged in the east before moving north.

According to the book, it was also visible in the west from May 24 to June 23 of that lunar month. However, according to calculations by several astronomers in the 1980s, the comet's closest approach to the sun occurred 240 B.C., between March 22 and May 25. These calculations also supported its apparent westward migration from the east. migration from east to north to west.

Therefore, it is frequently stated that March 30 is the most likely day for its initial observation, albeit it may not be the brightest. Astronomers in the Middle East, Asia, and finally, Europe monitored and documented every comet passage that followed. The comet, which appeared in 1066 and coincided with the Battle of Hastings, is depicted on the Bayeux Tapestry. Witnesses then reported seeing a comet four times the size of Venus.

For the Next 100 Million Years, the Footsteps on the Moon will Remain There

The absence of an atmosphere on the moon means that there is no wind to wear away the surface or water to wash the footprints away. Therefore, the Apollo astronauts' footprints, space, and rover prints will remain there for millions of years.

Because the moon has no atmosphere and all surface water is frozen as ice, there is no wind or water erosion on the moon like on Earth. Additionally, there isn't any volcanic activity on the moon to alter its surface features. Nothing is folded back inside, and nothing is washed away.

The Moon is subject to meteorite bombardment, which modifies the surface. One tiny space rock might quickly destroy a footprint on the moon. The solar wind, a stream of charged particles that originates from the sun, is also exposed to the moon because it lacks an atmosphere. Over time, this stream of charged particles behaves almost like weather on Earth to scour the moon's surfaces, but the process is extremely slow.

The Price of the NASA Space Suit is $12 Million

Suits for spacecraft travelers are now available from Final Frontier Design and cost between $125,000 and $150,000. Currently, the business is working on cases that can endure the forces of deep space and sell for less than $2 million each.

Since 2007, NASA has invested about $200 million in developing various space suits and technology. However, according to an Inspector General audit last year, they are not even entirely flight-ready. Of that sum, $12 million was spent on flight suits currently undergoing testing and will be worn by astronauts on Orion deep space missions in the 2020s.

In Space, there is Floating Water

Around 10 billion light years distant, astronomers discovered a gigantic water vapor cloud that contains 140 trillion times the volume of water in the oceans on Earth. It is the most significant water find ever made. Matt Bradford, a scientist at NASA's Jet Propulsion Laboratory in Pasadena, California, said that the atmosphere surrounding this quasar is unique because it creates this enormous mass of water. It serves as yet additional proof that water permeates the entire cosmos and has done so from the beginning of time. One of the teams that made Bradford lead the discovery.

There are 4,108 Exoplanets in the Universe

According to astronomers, our galaxy contains about one exoplanet for every star. So naturally, some stars have a lot of planets; for instance, our Sun has eight. Of course, some stars don't have any, too. But planet formation appears to be more common than uncommon if a star lives long enough.

However, this does not imply that astronomers can map all those billions of stars. The numbers may be smaller regarding exoplanets that have been quantified or tallied in some manner.

Currently, there are 4,108 confirmed exoplanets in the running total of known extrasolar planets.

The First Soft Drink Sipped in Space was Coco-Cola

When astronauts on the Space Shuttle Challenger drank from the Coca-Cola Space Can in 1985, it was one of the first soft beverages consumed in space. Finally, after more than 36 years, the company is bringing the enchantment of space to Earth for a brief period with the introduction of an extraterrestrial flavor and a galaxy of otherworldly, immersive digital experiences.

As the first limited-edition product from Coca-Cola Creations, Coca-Cola Starlight will be released this month in the United States. Through various unexpected beverages and packaging designs, culturally relevant expressions, and creative collaborations, these products will slake the thirst for the discovery of Gen Z.

Based on the limitless potential of space, Coca-Cola Starlight combines the classic Coca-Cola flavor with surprising additions, such as a crimson color and a chilling flavor that simulates traveling through space.

There Might Be Life On Mars

It's both fantastic and awful to look for life on Mars. On the one hand, the planet is a wasteland with wintertime lows of -153o C (-225o F) and an atmosphere that, to the extent that it exists, is just 1% as dense as Earth's and primarily made up of carbon dioxide. The Red Planet, though, wasn't always such a mess. It was surrounded by seas and oceans throughout the first billion years of its 4.5-billion-year existence and was shielded by a thick atmosphere. Its magnetic field eventually stopped functioning, allowing the solar wind to rip away the atmosphere and the water to disappear into space.

The first billion years, though, gave Mars ample time to cultivate at least microbial life, some of which may have perished and left chemical clues on the surface or perhaps retreated below to continue thriving in deep, warm aquifers.

There Might Be a Diamond-Covered Planet

The American Astronomical Society claims to have figured out the science of diamond planets in a scientific report published in 2020. According to studies undertaken by researchers at Arizona State University, stars with high carbon-to-oxygen ratios—where carbides predominate over silicates—are what bind diamond planets to their stars.

Scientists researched how silicon carbide performs in these environments by simulating these planets' severe pressure and temperature conditions. They observed that carbide silicon transforms into silica and diamond under pressures greater than 50GPA and temperatures up to 2500 degrees Kelvin. Pluto's largest moon, Charon, and WASP-12b, a Jupiter-like exoplanet discovered 1,200 light-years from Earth, are among the planets and moons that have been flagged as prospective diamond hotspots. Even closer to home, astronomers believe that Jupiter and Saturn frequently receive diamond showers.

In 3.75 Billion Years, the Milky Way and Andromeda Galaxies will Collide

The Andromeda and Milky Way galaxies interact gravitationally with one another, just as Earth's gravity might tug on a nearby asteroid, setting it on a collision course with our planet. The two galaxies are now descending toward one another at a speed of around 37 miles per second due to this (60 km per second). This cosmic disaster won't happen for 4.5 billion years because Andromeda is 2.5 million light-years away; 1 light-year is equal to 5.9 trillion miles (9.5 trillion km). Fortunately, the distance between stars makes it unlikely that anything will collide. However, the merger will alter the directions taken by stars inside each galaxy.

Two Identical Metal Pieces will Permanently Link Together If They Make Contact in Space

The oxygen in our atmosphere generates a thin film of oxidized metal on every exposed surface, preventing this from occurring on Earth. In addition, the oxidization layer is a handy barrier that prevents metal fragments from adhering to one another.

However, there is no oxidation layer in the absence of gravity in space. You suddenly have one continuous metal thing and a lot of explaining for your mission commander if the atoms of two metal objects come into contact with one another. We can carry naked chunks of metal to strike each other with and not worry about how inconveniently sticky they are, only in rare places like Earth with our highly caustic oxygen-rich atmosphere.

A Portion of the Earth Previously Orbited the Moon

The Earth and Moon are thought to have formed due to a massive collision between an early protoplanet and an astronomical entity known as Theia, according to the analysis of samples recovered from the NASA Apollo missions.

According to the capture theory, the Moon was once a roving body (similar to an asteroid) that formed somewhere in the solar system and was drawn in by the gravity of Earth as it went by. On the other hand, the accretion theory postulated that the Moon formed simultaneously with the Earth. Finally, the fission theory concluded that Earth had been spinning so quickly that some material had finally broken away and started to orbit the planet.

The Sun is Home to 99.86% of the Mass of Our Solar System

About 1,988,550,000,000,000,000,000 billion kilograms make up the mass of the Sun. which is around 330,000 Earths' worth. Helium makes up about 25% of, and hydrogen makes up about 73% of the mass of the Sun. The remaining components include heavier substances, including oxygen, carbon, neon, and iron.

Despite making up 99.86% of the mass of our Solar System, it is referred to as a yellow dwarf. This is because it dates back roughly 4.6 billion years. Like most other stars, it was created when a massive chemical cloud collapsed under the force of gravity into a core mass. This mass was eventually sufficiently compressed by gravity to reach a critical density and temperature, at which time nuclear fusion began in its core. Thus, our Sun was created.

Mercury and Venus Don't Have Moons

Our solar system is home to hundreds of moons; certain asteroids have even been discovered to have minor companion moons. Provisional moons are those that begin with a letter and a year. They will be given a proper name when more observations support their discoveries.

Mercury and Venus are the only terrestrial (rocky) planets in the inner solar system without any moons, while Earth has one and Mars has two minor moons. On the other hand, the ice giants Uranus and Neptune, as well as the gas giants Jupiter and Saturn, all contain several moons. With their strong gravitational fields, these planets could engulf smaller objects as they developed in the early solar system.

Mars Has a Volcano Three Times Bigger Than Everest

Mars has a three times bigger volcano than Everest, so it would have been a daunting place to investigate when it was young. Rivers of water and lava flowed across the surface of the Red Planet. At the time, Olympus Mons and the three summits of Tharsis Montes, a group of four volcanoes, were all rising to heights more significant than any mountain on Earth.

These are all stunning peaks. Olympus Mons, however, rises above the rest, measuring an astounding 16 miles (26 kilometers), or roughly three times the height of Mount Everest. As a result, Olympus Mons is now the largest volcano in the solar system. It is necessary to appreciate that Olympus Mons is more than just a tall volcano. It has a perimeter as well. The width of Olympus Mons is around 20 times that of its height. From edge to edge, its diameter stretches for 370 miles (600 kilometers).

Olympus Mons would cover the entire state of Arizona if it were placed atop the United States. Likewise, France will be covered if you spread it across Europe. According to a 2011 study, the volcano's volume is estimated to be about one million cubic miles (4 million cubic kilometers), which dwarfs everything on Earth. That is over 100 times the size of Mauna Loa, the planet's giant volcano.

Eighty-Three Moons of Saturn are Currently Known

Twenty other moons await confirmation of discovery and official designation, and the International Astronomical Union has confirmed and named 63 moons (IAU). The largest of Saturn's moons is the massive Titan, while the smallest is barely more significant than a football stadium. The moons of Saturn form, strengthen, and materialize the magnetosphere and rings. Titan, larger than Mercury, is just one example of a Saturnian moon among the largest in the Solar System. Others have distinctive qualities while being smaller than our Moon. So, let's look at Saturn's top five largest moons: Titan, Rhea, Iapetus, Dione, and Tethys.

Jupiter, a Gas Giant, is a Failed Star

Because it has the same ingredients as the Sun (hydrogen and helium), Jupiter is referred to as a failed star because it lacks the mass to have the internal pressure and temperature required to allow hydrogen to fuse to helium. This energy source drives the Sun and most other stars. We can't refer to the solar system as a double star because Jupiter only has 0.1 percent of the mass of the Sun and is, therefore, not a star.

Uranus Appears To Orbit the Sun Sideways

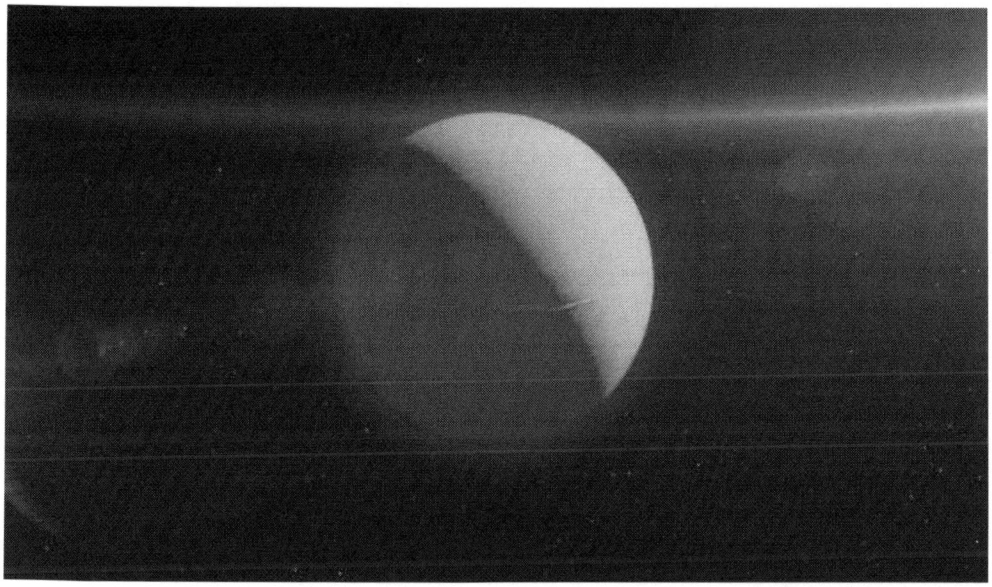

Uranus, the third-largest planet in our solar system and the seventh from the Sun, is extraordinarily chilly and windy. The ice giant rotates at a nearly 90-degree inclination from the plane of its orbit, encircled by 13 hazy rings and 27 tiny moons. Uranus appears to spin sideways while orbiting the Sun due to this unusual tilt.

Astronomer William Herschel discovered Uranus in 1781, the first planet to be located with a telescope, despite initially believing it to be either a comet or a star. Two years later, due to observations by astronomer Johann Elert Bode, the object was unanimously acknowledged as a new planet.

Pluto is No Longer Considered a Planet but a Dwarf Planet

The International Astronomical Union (IAU) degraded Pluto to a "dwarf planet" in August 2006. Only the outer Solar System's gas giants and the inner Solar System's rocky planets will be referred to as planets from this point on. The area of space that is smaller than the diameter of Jupiter's orbit around the sun is referred to as the "inner Solar System." Along with the terrestrial planets

Mercury, Venus, Earth, and Mars, it also encompasses the asteroid belt. Of course, Jupiter, Saturn, Neptune, and Uranus are the "gas giants." There used to be nine planets, but today there are just eight.

Pluto and Charon May Be Two Dwarf Planets in One System

Charon, Pluto's moon, is almost half its size. Because of the size of the moon, Pluto and Charon are frequently referred to as a twin dwarf planet system. They are 12,200 kilometers apart (19,640 km).

The Hubble Space Telescope captured photos of Pluto and Charon in 1994 when it was around

30 Astronomical Units (AU) from Earth. The distance between the Sun and the Earth's orbit is one AU, or 93 million miles or 150 million kilometers. These pictures showed that Pluto has a reddish tinge while Charon is a more neutral grey, indicating that their surfaces are composed and organized differently.

The Solar System Contains Five Officially Recognized Dwarf Planets

The International Astronomical Union is in charge of naming and categorizing celestial objects. Pluto, Ceres, Eris, Makemake, and Haumea are the five dwarf planets in our solar system that are formally recognized. Numerous additional celestial objects are being examined for inclusion in this category, and it is thought that the solar system may contain thousands of dwarf planets.

In our solar system, dwarf planets are celestial bodies roughly spherical, revolve around a star, and frequently contain comets, asteroids, or other dwarf planets nearby.

The Fifth-Largest Moon in the Solar System is Earth's Moon

The Moon on Earth is the sixth largest of the more than 200 moons that orbit planets in our solar system. Because no one was aware that there were any other moons until Galileo Galilei discovered four moons orbiting Jupiter in 1610. Earth's lone natural satellite is simply referred to as "the Moon."

The Lunar Reconnaissance Orbiter, the twin ARTEMIS spacecraft, and the upcoming Artemis program to return astronauts to the Moon are the three robotic spacecraft that NASA is currently using to explore the Moon.

Enceladus, the Moon of Saturn, is the Solar System Body with the Highest Reflectivity.

The Greek mythological titan for whom Enceladus is named. Even though this moon is small—only approximately 310 miles (500 kilometers) across—pictures from the Voyager spacecraft in the 1980s showed that its icy surface is surprisingly smooth in some parts and dazzling white all around.

Enceladus is the solar system body with the highest reflectivity. For many years, scientists had no explanation.

Enceladus' surface temperature is shallow, at roughly minus 330 degrees Fahrenheit since it reflects so much sunlight (minus 201 degrees Celsius). However, it is not as chilly and lifeless as it seems.

In Five Billion Years, Earth Will Be Swallowed Up By the Sun

The last stages, when the Sun will be roughly 12 billion years old, will see the most rapid phase of its evolution into a red giant. It will likely grow until it has a maximum radius of 1.2 AU, swallowing both Mercury and Venus (180 million km; 110 million mi). After that, Earth's orbital radius will get smaller due to tidal interactions with the Sun's outer atmosphere. In addition, the drag from the Sun's chromosphere would shorten Earth's orbit. These processes will cancel out the effects of the Sun's mass loss, and the Sun will probably swallow Earth in 7.59 billion years.

Massive Stars Explode in Supernovae To Finish Their Lives

Each star's core converts hydrogen into helium during the initial stages of formation. This first-stage nuclear reaction occurs in sun-like stars, red dwarfs slightly larger than Jupiter, and supermassive stars that are tens or hundreds of times as massive as our own. A star's core temperature rises with mass, so its nuclear fuel is used more quickly. When a star's core runs out of hydrogen to fuse, it compresses and heats up, at which point it may start combining even heavier elements if it becomes hot and dense enough. Sun-like stars will eventually become hot enough to fuse helium into carbon after hydrogen burning is finished, but that's it for the Sun. To proceed to the following phase, carbon fusion, you need a star that is eight (or more) times as massive as our Sun.

Uncrewed Spacecraft Have Been to Every Planet in the Solar System

Space exploration is the study of the reaches of the cosmos beyond Earth's atmosphere using crewed and uncrewed spacecraft to advance humankind's understanding of the cosmos. The section Chronology of crewed spaceflights has a comprehensive list of all crewed spaceflights, together with information on the accomplishments and crew of each mission.

People have always regarded the heavens and pondered the nature of the celestial bodies seen in the night sky. However, in the 20th century, technological advancements in electronics and other fields and the creation of rockets made it possible to launch machines, animals, and eventually humans into space.

For More Than Forty Years, Voyager 1 and 2 Have Been in Operation

On September 5, 1977, NASA's Voyager 1 probe launched, a few weeks after its twin, Voyager 2. The two Voyager spacecraft visited Jupiter, Saturn, Uranus, and Neptune during their epic "Grand Tour" of the solar system's giant planets.

However, their labor did not end there. Both spacecraft kept moving forward, pushing deeper into the unknown, chilly, and dark regions distant from the sun. 40 Epic Photos from NASA's Voyager Grand Tour

Then, on August 25, 2012, Voyager 1 broke loose and entered interstellar space, making history as the first artificial object to do so. According to mission crew members, Voyager 2, which traveled a different path through the solar system, will also likely leave the sun's sphere of influence in the coming years.

The Sphere-Shaped Oort Cloud Symbolizes the Solar System's Edge

Beyond Pluto and the most distant areas of the Kuiper Belt is where you'll find the Oort Cloud. The Oort Cloud is assumed to be a giant spherical shell that envelops the Sun, planets, and Kuiper Belt Objects. It is in contrast to the flat plane in which the planets of our solar system revolve around one another. It envelops the sun and planets in its orbit like a massive, dense bubble of ice and comet-like debris. In rare instances, the frozen bodies that make up the Oort Cloud may attain mountainous sizes.

Approximately 1.4 Billion Years Ago, a Day on Earth was Only 18 Hours and 41 Minutes

According to a report in The Guardian, recent research has shown that approximately 1.4 billion years ago, the amount of time it took for the Earth to complete a full rotation around its axis was approximately 18 hours and 41 minutes, rather than the more standard 24 hours. According to recent calculations presented in the Proceedings of the National Academy of Sciences, the length of a day on Earth is increasing at an average rate.0000135 seconds every year. This pattern is expected to continue millions of years into the foreseeable future. Researchers came to their conclusion regarding Earth's deep time when studying ancient sediment deposits for indicators of climatic change associated with periodic shifts in the planet's orbit. It led them to conclude that Earth has existed for an awfully long time.

A Space Station is About the Size of a Soccer Field

In celebration of the beginning of the 2014 World Cup in Brazil, NASA has created a graphic that compares the dimensions of the International Space Station (ISS) to those of a soccer field, also called a football field or a football pitch.

The station's weight, including its enormous solar arrays, is 924,739 pounds and takes up nearly the entire playing field area. As a result, the complex now boasts more usable space than a traditional six-bedroom house, a gymnasium, two baths, and a bay window that provides a panoramic view in all directions.

Since November 2, 2000, humans have continually occupied the space laboratory roughly the size of a football field and travel at a speed of around 17,500 miles per hour. This location is approximately 250 miles above the surface of the Earth.

Sky watchers of all ages can observe one of NASA's most significant accomplishments by venturing outside and taking a few steps backward and forwards. Viewing conditions for the International Space Station are excellent multiple times each month. Even though the International Space Station is constantly going around the planet, it is only sometimes possible to see it. Finding it requires a little advanced preparation, as well as clear skies.

The Sky on Mars is Reddish, while the Twilights are Blue

Iron oxide, which is similar to rust, is found in the soil of Mars, which is why it is known as the Red Planet. Even without a telescope, people on Earth can observe the peculiar ruddy tint that the planet possesses. The ancient Romans named it in honor of their God of War (Ares in Greek mythology).

To human viewers on the red planet, sunsets on Mars would have a bluish tint, similar to how colors on Earth become more vibrant when the Sun sets. However, when fine dust is in the air, the blue closer to the Sun appears considerably brighter. In contrast, when there is typical daylight, the rusty dust hue that is characteristic of Mars is more noticeable.

In Space, Astronauts Experience a Height Increase

When astronauts travel into space, their faces may give the appearance of being fuller or swelling out. This change may be more noticeable to observers.

Because of the Earth's gravitational pull, our hearts must put more effort into bringing blood to our brains and heads. However, because of the microgravity environment of the International Space Station (ISS), physiological fluids are no longer drawn down, and the heart has a much easier time pumping blood to the head, which results in an accumulation of fluids in the upper body and face.

After a few weeks, the facial swelling will typically begin to subside as the fluid levels in the body return to their normal levels.

Proxima Centauri is the Name Given to the Star System Located 4.25 Light Years Distant from Earth

A distance of slightly more than four light years separates our home planet from the star system Proxima Centauri, which may be found in the constellation Centaurus (The Centaur). Even though it looks highly luminous when observed via Hubble's telescope, Proxima Centauri cannot be seen with the human eye. This should be expected because the star is nearest to our solar system.

It has a brightness far lower than typical, and in relation to the masses of other stars, it is only around an eighth of the Sun's mass. Nevertheless, it elevates it to the status of a reasonably significant star.

In 1998, a Rose was Successfully Launched into Space

1998 marked the beginning of IFF's collaboration with the Wisconsin Center for Space Automation and Robotics (WCSAR), which was established as a NASA Commercial Space Center (CSC) at the University of Wisconsin. The mission of WCSAR is to assist private industry in exploring new product avenues in space. There are fifteen SCs across the country, and NASA's Space Product Development program at the Marshall Space Flight Center supports them. ASTROCULTURETM was the ideal solution for IFF's needs. On October 28, 1998, a miniature rose that IFF had chosen could embark on a journey into space on the shuttle Discovery for ten days (STS-95).

The Number of Known Exoplanets Has Increased to About 4,000 and will Likely Continue to Rise

Almost 4000 planets are orbiting faraway stars, yet not one provides the conditions necessary for life to originate on its surface. As a result, the Earth is an even more remarkable place than we had previously supposed.

Suppose the temperature at the surface of a planet is stable enough to sustain liquid water at that temperature. In that case, that planet is thought to have a reasonable possibility of harboring life. On the other hand, this depends on how closely it orbits its star; if it is too far away, any present water will turn to ice.

There are Auroras on Other Planets and Even On Some Moon in Our Solar System

Auroras will appear on any planet with a sufficiently dense atmosphere located in the path of the solar wind. For example, auroras on Venus have an irregular shape because the planet lacks a magnetic field. Auroras on other planets with magnetic axes significantly offset from their rotational axes have severely distorted forms of the auroral oval.

Auroras have been captured on camera on Jupiter and Saturn and the moons of some planets. However, because it lacks the necessary atmosphere, our moon does not have aurora-like other celestial bodies do.

There is a good chance that the same phenomena will occur on planets that are part of other solar systems. For example, when a star releases high-energy particles at the solar system's core, any planet in that system that has a sufficiently dense atmosphere will experience a phenomenon known as an aurora.

Space Debris That Enter the Atmosphere of Earth and Immediately Burn Up are Shooting Stars

Meteoroids are fragments of rock and dust in space that can be as small as dust grains or as large as minor asteroids. Therefore, consider them to be "rocks from space." Meteors are fireballs or "shooting stars" that form when meteoroids enter the atmosphere of Earth (or the atmosphere of another planet, such as Mars) at high speeds and burn up. The meteorite is the name given to a meteoroid that has survived its journey through the atmosphere and has arrived safely on Earth.

Orange Spacesuits are Less Common Than White Ones, But They Do Exist

Though most astronauts wear white spacesuits, the suit used to ascend and enter the space shuttle is called the Advanced Crew Escape Suit (ACES) and is orange. The color orange is referred to as "International Orange" and not any other variation. This specific color stands out quite clearly against the background of any kind of scenery, but notably in the water.

Extravehicular Activity (EVA) suits are characterized by increased bulk and white coloration. It is because white can reflect the intense heat of the sun. In addition, it stands out clearly against the inky backdrop of space.

Quasar 3C273 Has the Highest Temperature in Space

Quasar 3C273 is still one of the brightest and hottest quasars, and its distance from Earth is around 2.5 billion light-years. This quasar was the first to be positively identified as a quasar. According to a study conducted in 2016 and published in The Astrophysics Journal, the temperatures at the core of this quasar might reach more than 18 trillion degrees Fahrenheit (10 trillion Celsius).

The Boomerang Nebula is the Region of Space with the Lowest Temperature

The Boomerang Nebula is a juvenile planetary nebula that may be found in the direction of the constellation Boomerang, which is around 5,000 light-years away from our planet. It is called the Boomerang Nebula. In 1995, researchers discovered that this particular region of space has a temperature of just 1 K, approximately equivalent to a temperature of −457 degrees Fahrenheit (−272 C). Absolute zero, often known as 0 Kelvin, is a fictitious temperature that is the coldest physically feasible and at which all atomic and molecular activity would come to a complete and total halt. This temperature is the most out that can be imagined.

The Boomerang Nebula, shaped a bit more like a bowtie than a boomerang, is made up of gas released by a dying star in its core. Since the beginning of these last 1,500 years, the star has been suffering a decrease in mass equivalent to one-thousandth of a solar mass every year. And it is believed that the winds, which can reach speeds of up to 310,000 miles per hour (500,000 kilometers per hour), responsible for driving away this gas, are the source of the nebula's record-breaking low temperatures.

The Whirlpool Galaxy, Also Known as M51, was the First Spiral Structure in the Cosmos To Be Discovered

Long, winding paths of stars and gas, liberally sprinkled with a great deal of space dust, make up the gigantic arms that spiral out from the center of the Whirlpool Galaxy. These arms function as factories for developing new star clusters by compressing hydrogen gas and producing new star clusters.

A Year's Worth of Travel at the Speed of Light is Measured in Terms of a Unit Called a Light-Year

The distance that light can travel in a single year is expressed in a unit of measurement called a light year. The speed at which light travels through interstellar space is approximately 186,000 miles (300,000 kilometers) per second. The distance that light can traverse in one year is around 5.88 trillion miles (9.46 trillion kilometers). The immense lengths of freedom are typically measured in terms of light-time. It refers to the amount of space traversed by light in a given time.

Light moves very quickly; nothing can move faster than light. For example, Jupiter is approximately 484 million miles away. Thus, it takes the sun's rays 43.2 minutes to reach the planet. Although light travels at the speed of light, the distances are enormous. Light is capable of traveling 671 million miles in just one hour. The distance between the Sun and Earth is approximately eight minutes of light.

The Planet Pluto Has Mountains of Its Own

Since Pluto is a relatively small world located on the solar system's outskirts, scientists anticipated its surface would be pockmarked with numerous craters. However, this was no longer the case after the New Horizons spacecraft from NASA flew past Pluto in 2015 and sent back images that irreversibly changed our perspective of the planet.

Ice mountains measuring 3,300 meters (11,000 feet) in height have been found on Pluto, which indicates that the planet's geology must have been active as recently as 100 million years ago. This is one of the most astonishing discoveries ever made. Geological activity, however, requires a source of energy, and the location of that energy source within Pluto remains a mystery.

Venus in Our Solar System Has the Most Extraordinary Average Surface Temperature of Any Planet

Because of its thick atmosphere, which traps heat and creates a greenhouse effect that can get out of control, Venus has surface temperatures that are hot enough to melt lead, making it the hottest planet in our solar system. The presence of the greenhouse effect causes the surface temperature of Venus to be roughly 390 degrees Celsius (nearly 700 degrees Fahrenheit), a greater temperature than it would be if the greenhouse effect did not exist.

On Venus, a Day Lasts for More Than the Equivalent of an Entire Year

One day is the amount of time required for a planet to complete one revolution around its axis, while one year is the amount of time needed for a planet to complete one orbit around the Sun. Venus is one of only two planets in our solar system that revolves around itself in a clockwise direction. As a result, its rotational speed is significantly lower than that of the other planets. Some people believe this occurred because it was jostled in a different order by another planet. Others think that it gradually slowed down until it stopped moving and then started turning in the opposite direction. As a direct result, it takes Venus 243 Earth days to complete one spin around the Sun. On the other hand, it only takes Venus 225 Earth days to orbit the Sun. As a direct result, the length of one day on Venus is equal to more than 365 days on Earth.

Pluto isn't the Last Planet in the Solar System

The Oort cloud can be considered the equivalent of Terra Australis in the Solar System. It is a speculative bubble of icy debris that scientists have predicted to be the most distant part of our Solar System, but it has not yet been discovered. It makes it an exciting undiscovered mystery. For example, in a scenario where the distance between the Earth and the Sun was reduced to one centimeter, the Oort cloud would be located 500 kilometers away. Despite this, there are regions of the far reaches of the Solar System that we know of, such as the Kuiper Belt.

Uranus Spins Sideways

According to the guide that NASA has created for Uranus, the planet rotates on its side, giving the impression that it is rolling around the sun like a ball. This is because the planet may have been involved in a catastrophic collision in its distant past. This theory best explains its peculiar orientation, which places it at an angle of approximately ninety degrees concerning the other planets.

NASA considers the seasons on Uranus to be the most intense of anywhere in the solar system because of the planet's tilt. The sun is positioned to shine directly over Uranus's north or south pole for approximately one-quarter of each Uranian year (or 21 Earth years, given that one Uranian year lasts 84 years). This means that for more than twenty decades here on Earth, half of Uranus will never, under any circumstances, be exposed to sunlight. Scientists monitor the severe seasons on Uranus, and they anticipated that the planet's equinox in 2007 might result in peculiar weather conditions.

The Moon Io Orbiting Jupiter is Home to Towering Volcanic Explosions

The tumultuous nature of Jupiter's moon, Io, may come as a shock to those who are accustomed to the tranquility of Earth's moon. There are hundreds of volcanoes on the moon of Jupiter, making it the most active moon in the solar system. These volcanoes release sulfuric gas plumes that can travel as far as 190 miles (300 kilometers) into the moon's atmosphere. Io's volcanoes are responsible for releasing one ton (more than 900 kilograms) of gases and particles into space close to Jupiter every second, according to a statement from NASA (opens in new tab).

The tremendous forces to which Io is subjected due to its location within Jupiter's gravitational well and magnetic field cause the moon's volcanic behavior. In addition, the tension and relaxation of the moon's interior, which occurs when the moon orbits closer to and further away from the planet, respectively, produces enough energy for volcanic activity.

However, because scientists have not yet figured out how heat travels through Io's interior, it is difficult to anticipate where volcanoes exist based solely on scientific models.

Chapter 2: Amazing Facts About Stars and Constellations

"In this single galaxy of ours, there are eighty seven thousand million suns." - **Arthur C. Clarke**

The Nearest Star is the Sun

It is reasonably incredible to consider that our own Sun, which is only 150 million kilometers from Earth, is an average example of all the stars in the Universe. Our own Sun is a G2 yellow dwarf star currently in its main sequence phase of existence. Since 4.5 billion years ago, the Sun has been happily turning hydrogen into helium at its core, and it will probably do so for at least another 7 billion years. The Sun will swell to several times its current size when its fuel runs out, turning it into a crimson monster. The Sun will inevitably swallow Earth, Venus, and Mercury as it enlarges.

The Same Material is Used to Make Stars

Cold molecular hydrogen clouds gravitationally collapsed to form all stars. When a cloud collapses, it splits into several parts that will eventually coalesce into individual stars. The material gathers into a ball that keeps collapsing due to its gravity until nuclear fusion may start in the center of the ball. This primordial gas, which constantly contains roughly 74% hydrogen and 25% helium, was created during the Big Bang. The ratio in our Sun is closer to 70% hydrogen and 29% helium. However, all stars begin with additional trace elements and a mixture of 3/4 hydrogen and 1/4 helium.

The Stars are Evenly Balanced

You may not be aware, but stars are constantly at odds with themselves. A star is being drawn inward by the combined gravity of all its mass. The star would keep crashing for millions of years until it reached its minimum size—possibly as a neutron star—if there was nothing to stop it.

However, light exerts a force that resists the star's gravitational collapse. A star's core undergoes nuclear fusion, producing tremendous energy. The photons push outward as they travel from the star's center to the surface; this trip could take 100,000 years. Red giants are formed when stars expand outward as they get brighter. And when the light pressure runs out, they disintegrate into white dwarfs.

Red Dwarf Stars Make Up Most Stars

Red dwarfs would be the most extensive collection of stars if you could gather all the stars into one place. These are stars with less mass than the Sun, by less than 50%. Red dwarfs can even have groups that are only 7.5% as small as the Sun. Below that, the star lacks the gravitational pressure to increase its core temperature and initiate nuclear fusion. These are referred to as failed stars or brown dwarfs. Red dwarfs can consume their fuel for 10 trillion years before running out of hydrogen because they burn with less than 1/10,000th of the energy of the Sun.

Color = Temperature x Mass

Stars can be any hue, from red to white to blue. A red star has a Kelvin temperature of less than 3,500, making it the coolest hue. Like our Sun, stars have an average temperature of 6,000 Kelvin and are yellowish-white. Blue stars are the hottest stars and have surface temperatures of 12,000 Kelvin. Therefore, a star's temperature and color are related.

A star's temperature is determined by its mass. The star's core will be larger and more capable of performing nuclear fusion as mass increases. This implies that the star's surface receives more energy, which raises the star's temperature. Red giants are a challenging exception to this rule. The mass of our Sun can be found in a normal red giant star, which has always been white. But as its life draws to a close, its luminosity increases by a factor of 1000, giving the impression that it is unusually brilliant. A giant blue star, however, is only large, massive, and hot.

The Majority of Stars are Multiples

Even though there may be only stars in the sky, many come in pairs. These are binary stars, where two revolve around a single gravitational pole. Other systems exist that have 3, 4, or even more stars. Imagine the breathtaking sunrises you would see if you woke up on a planet with four stars orbiting it.

The Most Prominent Stars would Consume Saturn

Some monstrous stars out there truly make our Sun appear little. Betelgeuse, a famous red supergiant, is located in the constellation of Orion. It is 1,000 times larger but has a mass roughly 20 times that of the Sun. But it's not much. On the other hand, monster VY Canis Majoris is the biggest known star. This star is estimated to be 1,800 times bigger than the Sun and would completely engulf Saturn's orbit!

The Stars That Have the Most Significant Mass Also Have the Shortest Lifespan

As indicated above, low-mass red dwarf stars can burn through all of their fuel for 10 trillion years before they run out. So, about the most prominent stars, we are aware of. The exact opposite, though, is true. These giants are highly active and can have 150 times the mass of the Sun. One example is Eta Carinae, one of the most massive stars we know of, about 8,000 light-years away.

This star has 150 solar masses and is thought to release 4 million times as much radiation.

In contrast to our own Sun, which has quietly burnt away for billions of years and will continue to do so for billions more, Eta Carinae has likely only lived for a few million years. Eta Carinae is also predicted to go supernova at any time, according to astronomers. So, you could read from it at night and see it during the day since it would be so brilliant.

There are Countless Stars

The fact that our galaxy has 200–400 billion stars may astonish you. Each is a unique island in space, maybe with planets, and some might even contain life. The Universe may include as many as 500 billion galaxies, each containing as many stars as the Milky Way or more. When you combine the two figures, you can see that the Universe may have as many as 2 x 1023 stars. That's 200,000,000,000,000,000,000,000.

Stars are Far Away

It's astonishing to think about the immense distances involved when there are so many stars in the Universe. For example, Proxima Centauri, a star that is 4.2 light-years from Earth, is the nearest star to it. In other words, the trip of light itself from Earth takes more than four years. It would still take more than 70,000 years to travel there from here, even if you tried to take the fastest spacecraft launched from Earth. It's just not possible right now to travel between the stars.

Galaxies are Either Elliptical, Spiral, or Irregular

The following illustrates three distinct varieties of galaxies, each of which may be identified by their shape. These are classified as elliptical, spiral, or irregular galaxies. Each type of galaxy—elliptical, spiral, and rare—holds its unique collection of stars, interstellar dust, planetary systems, and gaseous elements. These three distinct kinds of galaxies are all created in the same way, which is to say, by the action of gravitational forces. Galaxies with an elliptical form resemble spheres or cucumbers and are home to many elderly stars. Spiral galaxies are characterized by the presence of young stars and by the presence of a core disk that is ringed by spiraling arms. Galaxies considered irregular lack a discernible shape and contain old and young star formations.

The Majority of Galaxies Include a Central Black Hole at Their Centers

Possibly, the mass of a supermassive black hole is hundreds, thousands, millions, or even billions of times bigger than that of the Sun. The gravitational collapse of various astronomical phenomena collectively referred to as black holes have resulted in spherical areas of space from which nothing, not even light, can escape. These space areas do not allow for any kind of exit or escape. According to the evidence accumulated through observations, a supermassive black hole will likely be found at the center of practically every giant galaxy. For example, the Milky Way's galactic core has a supermassive black hole corresponding to the radio source Sagittarius A.

Distinct Colors Truly Exist Among the Stars

First, it's important to note that distinct colors truly exist among the stars. It was known in the 19th century and earlier that heating an object causes it to glow, and it was also recognized that the color of the glow changes depending on the object's temperature. This was true even before the 19th century. A man named Wien (pronounced "Veen"; he was German) was even able to apply some mathematics to this and compute the temperature of an object based on its hue, although it remained unclear why this worked. When an additional scientist named Planck contributed to the development of quantum mechanics in the 20th century, this was finally grasped.

Stars Have Atmospheres, Just Like Planets

Planets and stars are the celestial bodies most commonly thought of as having atmospheres; nevertheless, comets, which only have transient atmospheres, are another type of celestial body that can have atmospheres. One of two ways to obtain an atmosphere is for it to be accreted onto a celestial body through gravitational attraction; the other way is to be expelled from the celestial body through a process known as outgassing. Both of these methods are possible.

The mass of a celestial body and the temperature of its atmosphere both have a role in determining whether or not the body keeps its atmosphere. Generally, the greater the mass of the body and the lower the temperature of its atmosphere, the greater the likelihood that the body will be able to retain its atmosphere. For instance, a planet near a star that receives a great deal of radiation may eventually lose its atmosphere, as was the case with Mercury in our solar system.

On Average, an Atomic Nucleus's Density Can Be Roughly Analogous to the Thickness of a Regular Neutron Star

If a big star goes through a supernova and then collapses under its own gravity, a neutron star may be the type of stellar remnant left behind. The stars in the night sky, known as neutron stars, have managed to cram the most mass into the smallest space. Even though their radii are just around 12–13 kilometers (about 6 miles) long, their masses are likely many times that of the Sun. Their radii are only about 12–13 kilometers (about 6 miles) long. When viewed with the naked eye, neutron stars most definitely have a white coloration to them.

The name "neutron star" comes from the fact that virtually all of the star's constituents are neutrons, subatomic particles that do not carry a net electrical charge and have a mass that is somewhat greater than protons. Neutrons have a group that is slightly greater than that protons. Neutron stars are protected from further collapse by force referred to as quantum degeneracy pressure, which also contributes to the unusually high temperatures they display. The Pauli exclusion principle offers a rationalization for the occurrence of the phenomena that are responsible for the presence of this force. According to this theory, it is impossible for two neutrons or any other fermionic particle to exist at the exact location while still maintaining the same quantum state. This is true even if the site is in the same dimension. This is a fundamental principle that underpins the Copenhagen interpretation of quantum mechanics.

A neutron star is defined by having a surface temperature of approximately six times 105 degrees Kelvin and a mass that ranges from around 1.4 to about two solar groupings. The surface temperature of a neutron star is determined by its group. Neutron stars have an average density that ranges from 3.71017 to 5.91017 kg/m3, which is 2.61014 to 4.11014 times the thickness of the Sun. This density is practically identical to the density of an atomic nucleus, around 31017 kg/m3.

There is Roughly the Same Number of Binary Star Systems as Single-Star Systems

Scientists have found that binary star systems offer the most accurate technique by which to measure the mass of a star. In addition, astronomers can identify properties like temperature and radius based on the size, which can be calculated when the pair pulls on each other and causes it to become more prominent. These aspects contribute to categorizing solitary main-sequence stars across the cosmos.

A direct influence on life can come from a star in more than one solar system. Numerous planets that orbit a variety of different stars have previously been discovered. The path these stars take can impact the development of life, which requires a generally stable system to flourish. Even though binary and multiple star systems initially appear intimidating, specific star systems, such as wide and close binaries, could produce conditions under which life could eventually evolve. It is because one or more stars constantly move closer and farther away from the planets, which alters the amount of light, heat, and radiation the worlds are exposed to.

The Sky is Home To 88 Different Constellations

The sky is home to 88 different constellations, all created by humans by combining easily discernible star patterns. However, these stars can be very distant and do not communicate with one another in any way. Yet, looking at them from the ground, it appears they are situated next to one another.

Most northern constellations officially recognized today were initially constructed by ancient Sumerians and Babylonians and later by Greeks. Because of this, numerous stars and constellations have been linked to various old stories and legends. For instance, the constellation Lyra is symbolic of the celestial harp that was fashioned by Hermes and presented to Orpheus by Apollo. People and animals alike were mesmerized whenever Orpheus sang his adored bride Eurydice love songs during their courtship. Orpheus's mesmerizing lyre music persuaded the ruler of the underworld to let Eurydice return to the surface, but only on the condition that he did not look back at her as she followed him out of the underworld and into the light. Eurydice had died unexpectedly. Incapable of bearing the uncertainty that she was not following him, he turned around and saw Eurydice dissipate into Hades' domain after glancing in that direction. In remembrance of

Orpheus's heartbreakingly beautiful music and the woman he had to leave behind, Zeus hung Orpheus's lyre among the constellations after he left.

Sirius is the Brightest Star Seen in the Night Sky

Certain magnificent heavenly lights shine brighter than others, and the night sky may be a lovely place filled with stars. However, some stars shine brighter than others. Even on a night with clear skies, people living in cities may only be able to see the brightest stars against the city's light pollution.

On the other hand, if you go somewhere incredibly dark, you might see thousands upon thousands of stars twinkling above you. But even under that scenario, some stars would be substantially brighter than others, and the brightest stars will be the same select handful that people living in cities can see.

UY Scuti is the Most Massive Star Known to Humankind

According to Astronomy Magazine, German astronomers working at the Bonn Observatory in 1860 were the first to catalog UY Scuti. At the time, they referred to it as BD -12 5055. As a result of a second observation, scientists found that it varies in brightness over 740 days, which led to the variable star classification being assigned to it. The star is approximately 9,500 light-years away from Earth and is located near the core of the Milky Way galaxy. A hypergiant star that can be found in the Scutum constellation is known as UY Scuti. Hypergiant stars are scarce and shine with an incredible amount of light. They are more prominent than supergiants and giants combined. However, they lose a significant portion of their mass due to the swift movement of stellar winds.

47 Ursa Major is the Name of the Star in the Constellation Ursa Major

With an apparent magnitude of +5.03, it is visible to the human eye when conditions are favorable. The star 47 Ursa Major has a mass comparable to our Sun's. It has roughly 110% of the abundance of iron that the Sun does and is, therefore, slightly more metal-rich than the Sun. It has the same spectral class as the G1V star, the Sun, although it is approximately 5,855 K hotter than the Sun. Because 47 Ursae Majoris has an absolute magnitude of +4.29, its visual brightness is roughly 60 percent more than the Sun's.

In its core, 47 Ursae Majoris undergoes the same process as the Sun, known as nuclear fusion, which converts hydrogen into helium. Although the star's chromospheric activity suggests that it is approximately 6,000 million years old, evolutionary models point to a far more senior age of about 8,700 million years for the star.

Stars Die

Depleting a star's nuclear fuel ultimately results in the star's destruction. The star's mass is the determining factor in what happens to it after it has existed. Highly massive stars use hydrogen fuel rather rapidly; their temperatures are high enough to fuse heavier elements such as helium and carbon. When nothing is left to fuel the star, it will collapse, and what is known as a "supernova" will occur when the outer layers of the starburst. Assuming there is sufficient mass, what is left behind after a star's supernova explosion is either a "neutron star," which is the star's core after it has collapsed, or a black hole. A neutron star is the more stable of the two possibilities.

BPM 37093 is Fifty Light-Years from Earth

It is the largest diamond in the universe, and Lucy also was formed from the crystallization of a white dwarf. BPM 37093 A white dwarf star located in the constellation Centaurus is relatively close to us. It is believed that the atoms in the Moon-sized carbon core were forced so together by gravitational pressure that they stripped each other of the electrons from around their positively charged nuclei, and the mutual repulsion between the nuclei forced them into a crystalline lattice structure. It comprises a thin layer of hydrogen atmosphere resting on a carbon core roughly the size of the Moon. In the core, the atoms are believed to have been pushed so close together that they stripped each other of the electron. In February of 2004, researchers at the Harvard-Smithsonian Center for Astrophysics announced that new observations of the star BPM 37093 had provided conclusive evidence that it was, in fact, a crystalline white dwarf. It makes BPM 37093 the first object of its kind to be found. The white dwarf BPM 37093 is a pulsing star, and measurements of the pulsations show that the inside of the white dwarf is certainly solid crystalline carbon. This information was gleaned from observations of the pulsations. The unofficial name for BPM 37093 is "Lucy," and it comes from the Beatles song "Lucy in the Sky with Diamonds," which is a reference to the fact that BPM 37093 is essentially a massive diamond that exists in space. mv = 13, spectral type = DA V, diameter = approximately 4000 kilometers, mass = 1.14 times that of the sun, and distance = 15.3 parsecs.

Some Stars in Galaxies Do Not Circle Around the Center of Their Galaxy

Some stars in galaxies do not circle the center of their galaxy. All of the stars that have been found are extremely brilliant red giants. So, it is possible that in addition to the galaxies that Hubble discovered, the cluster also contains many fainter stars.

How can you know that this collection of stars does not belong to a galaxy? You can observe their motion as well as the motion of any other galaxies that are nearby. The observations demonstrate that the whole cluster's gravitational field, and not the attraction of any particular galaxy, is responsible for determining the motion of the individual stars inside the group. To put it another way, they are actually "intergalactic," which means they exist across galaxies.

Nebulae, which are Enormous Dust Clouds, are Stars' Birthplace

Nebulae, which are enormous dust clouds, are stars' birthplace. The effect of gravity is to cause the dust to clump together. As more and more dust gathers together, the gravitational pull becomes more significant, and the object begins to heat up, eventually evolving into a protostar. As soon as the core reaches the necessary temperature, nuclear fusion will start, and a new star will be born.

Main Sequence

Once something has become a star, it will maintain its status as a star by continuing to glow and release energy for billions of years. This phase of the star's evolution, known as the "main sequence," accounts for the vast bulk of the star's lifespan. During this time, a balance is achieved between gravity's desire for the lead to shrink and heat's desire for the star to get larger. The star will continue to exist in this state until it exhausts its hydrogen supply.

Giant Crimson Star

A fading star that has reached the very end of its growth into a star is known as a red giant. In around five billion years, our sun will transition from a yellow dwarf to a red giant, engulfing the inner planets and probably even Earth. Red giant stars expand to a diameter of 100 million to 1 billion kilometers (100 million to 620 million miles), one hundred to one thousand times wider than our sun's current size. The surface temperatures of these stars are quite low, reaching only 4,000 to 5,800 degrees Fahrenheit (2,200 to 3,200 degrees Celsius), which is a little more than half as hot as the sun's temperature. This is because the energy produced by these stars is distributed over such a large area. Because of this temperature change, stars begin to light in the redder section of the spectrum, earning them the term "red giant," although they typically appear more orange.

A Fleeting Glimpse of a Star

The effect that the Earth's atmosphere has on light refraction gives the night sky the appearance that the stars are blinking or moving about. When starlight travels through space and enters the atmosphere, it is exposed to the impact of the atmosphere's winds and regions that vary in temperature and density. This causes the starlight to take on different appearances depending on where it is. This causes the light from the star to flicker when seen from the earth below. When a ray of light travels from one medium to another, a phenomenon known as "bending" occurs. This process can be described quite well using the term "refraction." If it travels from a rare media to a dense media, it will bend toward the normal, but if it travels from a dense media to a rarer media, it will bend away from the usual.

Brown Dwarf

Brown dwarfs are astronomical objects that have a size that falls somewhere in the range between that of a massive planet like Jupiter and that of a relatively modest star. The vast majority of astronomers agree that a brown dwarf appropriately describes any object with a mass that falls somewhere between 15 and 75 times that of Jupiter. Because a thing with that mass range would not have been able to maintain the fusion of hydrogen like a typical star, many scientists refer to brown dwarfs as "failed stars."

Since 1995, astronomers could identify a few brown dwarfs nearby. Every single brown dwarf that has been found up to this point is a component of a binary system. A binary system is a system that consists of two stars orbiting around each other in opposite directions (just like the planets of our solar system orbit our star, the Sun).

Star Shape

A regular star form is a self-intersecting equilateral and equiangular polygon known by its other name, a regular star polygon. A typical star polygon is another term for a traditional star form. It was made by attaching one vertex of a plain, ordinary, n-sided polygon to another vertex that was not close to it and continue this procedure until the initial vertex was reached. This generated the item in question.

If we want to create a five-pointed star, for instance, we will need to draw a line from the first vertex to the third vertex, from the third vertex to the fifth vertex, from the fifth vertex to the second vertex, from the second vertex to the fourth vertex, and from the fourth vertex to the first vertex. This will complete the formation of the star. Using this method, we can fashion the form of a star.

Protostar

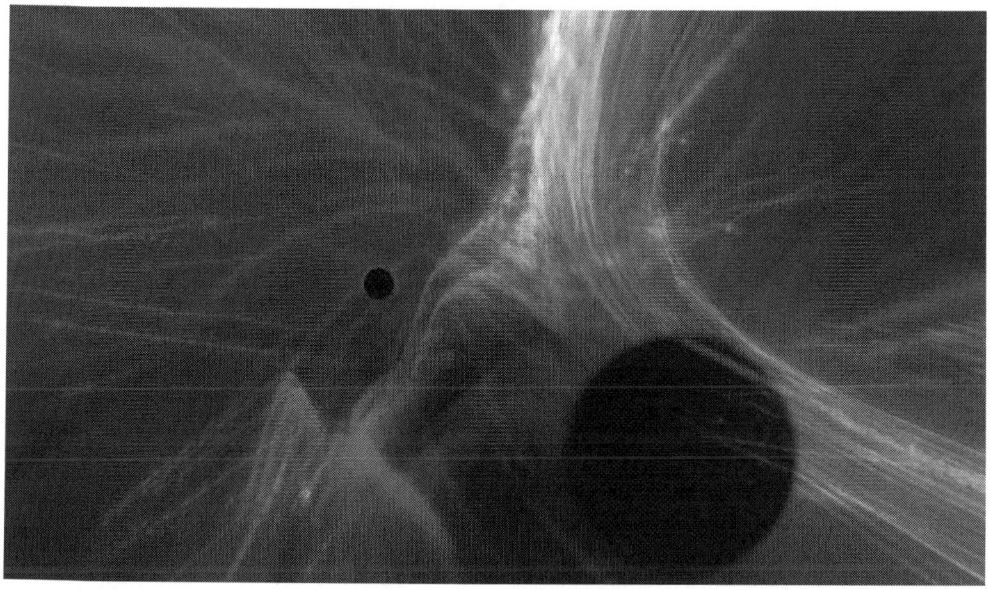

A young star that has not yet depleted the molecular gas in its surroundings due to gravitational contraction is referred to as a protostar. The core of a protostar has not yet reached the density and temperature necessary for the collision and fusion of hydrogen nuclei, which is required to form helium nuclei.

Heat is generated whenever the individual gas particles that make up the molecular cloud collide with one another. Consequently, a heated clump of molecules that came to be known as the protostar was produced. Infrared imaging makes it possible to see the formation of protostars, which may be distinguished from other components of the molecular cloud by their higher temperatures. In addition, depending on the size of the cloud in which they are created, a single cloud may give rise to more than one protostar.

T-Tauri Star

When the components falling into the Protostar stop falling and massive amounts of energy are released, a T-Tauri star will begin to form. Fusion reactions can't take place in the Tauri star's core because the average temperature there isn't high enough. The T-Tauri star has a lifespan of around

100 million years, after which it transitions into the Main sequence phase, the development phase that lasts the longest.

Giant Molecular Clouds

Our galaxy has a star formation rate that ranges from one to four solar masses of stars being created every single year. In contrast to elliptical galaxies, which are almost entirely barren of new star formation, spiral galaxies still have molecular gas reservoirs, which fuel new star formation. As a result, star formation is still occurring in spiral galaxies. In addition to the stars that can be seen so plainly from our home planet, the discs of spiral galaxies are made up of gas (the interstellar medium, ISM). These frigid, dense molecular areas are known as molecular clouds and are the birthplace of young stars. As this gas accumulates, it forms these molecular clouds. As a result, most star creation occurs in enormous molecular clouds, also referred to as giant molecular clouds (GMCs).

However, while we have a good understanding of how individual stars form, there is less consensus on how their natal clouds of gas accumulate, how long these clouds last, how star formation progresses for their lifetime, and how it has gone for the Milky Way's lifetime. This is although we have a good understanding of how individual stars form.

Implosions Cause Supernovas and White Dwarfs

As the star expands, the helium molecules at its core fuse together. The energy produced by this reaction keeps the core from crumbling in on itself. As soon as the helium fusion process is complete, the center will begin to contract and start fusing carbon. This technique is repeated repeatedly until the iron is found in the core. The collapse of the middle is a direct result of the iron fusion event, which consumes energy. As a result of this implosion, big stars transform supernovas, while lesser stars, like our sun, change into white dwarfs.

Chapter 3: Amazing Space Facts About Constellations

"The history of astronomy is a history of receding horizons." - **Edwin Powell Hubble**

The Latin Phrase "Set With Stars" is Where the Word "Constellation" Originates

The term "constellation," which was first used in the 14th century, may also be found in Middle English, where the phrase "set of stars" can be found. The term "constellation" was the first time it was used. The word "constellation" originates from the Late Latin word "Castelli," which means "set." The word "constellation" was borrowed from English. It is a group of stars that have been given a name due to the unique shape that they form in the sky. This distinctive shape has led to the naming of the constellation. This one-of-a-kind form can be easily identified and referred to by its specific moniker.

In most cases, they are named after figures from mythology, real people, animals, items from the real world, or inanimate objects. Sometimes, though, they are named after inanimate objects. Using the same clusters of bright stars, people from different parts of the world have produced a variety of distinct geometric patterns using their creativity. These patterns can be discovered in many other parts of the world.

Modern Applications May Be Obtained for the Constellations

There is also a modern application that may be obtained for the constellations. They are the ones who choose what names should be given to the stars. One of the most enjoyable aspects of going to conferences for astronomers is having the opportunity to discuss their work with colleagues from other institutions. And in most instances, they will want to describe to another person the specific stars or objects they are observing. If they merely provide the coordinates (numbers) for the star, the other person is unlikely to have a passing sense of where the star is located in the sky, even if they provide all the coordinates.

If, on the other hand, you tell someone that the star's name is Alpha Tau, then you will be aware that it is the brightest star in the Taurus constellation. The terms of the star systems in which they are contained, known as constellations, are where the names of the stars come from (all stars are in some constellation). The names, which advance in order from brightest to dimmest star in brightness, are indicated using Greek letters. For example, the star Beta Ori is the second most brilliant in the constellation Orion (also called Rigel).

Farmers Used Constellations for the First Time

One of the more useful applications of constellations was in agriculture. Before there were accurate calendars, people had no method of knowing when to plant or harvest their crops other than by seeing the constellations in the sky. In addition, the patterns of the stars might be more easily remembered thanks to the constellations. Ancient people, for instance, were aware that when the constellation Orion began to become completely visible, it was only a matter of time before winter arrived. They might also look at the Summer Triangle to determine when Spring and Summer were about to come. The patterns in the sky could more easily be recognized and interpreted because of constellations, which made it easier for farmers to make long-term plans and form agriculture.

How Stellations Work

The patterns in the sky could more easily be recognized and interpreted because of constellations, which made it easier for farmers to make long-term plans and form agriculture. In addition to this, navigating was made more accessible by the constellations. For example, finding Polaris, also known as the North Star (Little Dipper constellation), shouldn't be too challenging once you've located Ursa Minor.

A Religious Constellation or Astronomical Pattern

Religious use was likely the initial application of constellations. People believed that the gods resided in the skies and were responsible for their formation. In many cultures, people thought that the constellations in the sky were their gods trying to tell them a story. Because of this, identifying patterns in the sky, giving those patterns names, and making up stories about them seemed perfectly natural.

The Greeks were the ones who gave us the terms of our many constellations. And they gave the names of the constellations based on their legendary figures and heroes from their mythology. Therefore, there is a tale concealed within every constellation. For instance, the ancient Greeks believed Orion to be an accomplished hunter. It is said that Neptune was his father (god of the sea). On the other hand, the Egyptians believed that these very same stars represented Osiris. Every culture comes up with its very own way of understanding things.

The Stars are Divided into Many Groupings

The stars are divided into many groupings. Andromeda, Aquila, Auriga, Boötes, Cassiopeia, Cepheus, Corona Borealis, Cygnus, Delphinus, Draco, Equuleus, Hercules, Lyra, Ophiuchus, Pegasus, Perseus, Sagitta, Serpens, Triangulum, and Ursa Major, and Ursa Minor are the 21 Northern Constellations.

There are 15 Constellations Visible in the Southern Hemisphere

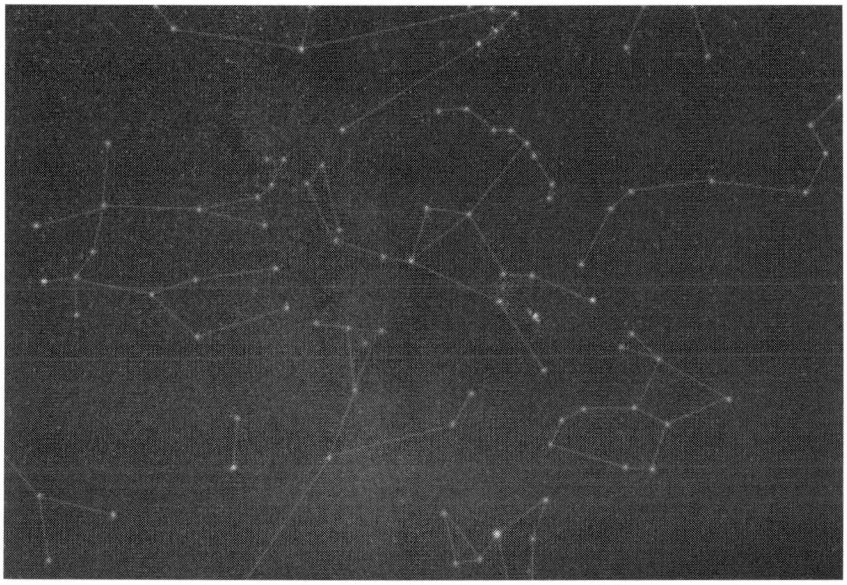

There are 15 constellations visible in the southern hemisphere: Ara, Argo Navis, Canis Major, Canis Minor, Centaurus, Cetus, Corona Australis, Corvus, Crater, Eridanus, Hydra, Lepus, Lupus, Orion, and Pisces Austrinus.

There are 12 Zodiacal Constellations in Astrology

There are 12 Zodiacal Constellations in astrology: Aries, Aquarius, Cancer, Capricornus, Gemini, Leo, Libra, Pisces, Sagittarius, Scorpius, Taurus, and Virgo.

Casper Vopel Created Two More Constellations in 1536

Until a German globe builder, Casper Vopel, created two more constellations in 1536, these original constellations were the only ones given names. Coma Berenices and Antinous are their names.

Dutch Astronomer Plancius Named Constellations

Until 1589, when a Dutch astronomer named Plancius started filling the southern celestial blank, there were no constellations in Ancient Greek with names for the stars in the south. Crux and Triangulus Antarcticus were his creations. He continued to design numerous such constellations after this.

Stars Employed to Pinpoint the Locations of Seagoing Ships

Before fancy navigational instruments aboard seagoing ships, the stars were employed to pinpoint their locations; latitude (North/South) was determined by the height of Polaris in the sky, and Ursa Minor (the Little Dipper constellation) was used to determine longitude (North/South).

Nakshatra is the Name for the Lunar Mansion in Hindu Culture

The Nakshatra is the name for the lunar mansion in Hindu culture. One of the 27 (and occasionally 28) sectors along the ecliptic is known as a nakshatra. Their names are associated with the most noticeable star patterns in the corresponding sectors.

Greek Mythology and Origin Tales are Associated with Each Constellation

Andromeda was the name of an Ethiopian princess, and the constellation that now bears her name was named for her. Andromeda was the daughter of Cepheus and Cassiopeia, who was bound to a rock as a sacrifice to the monster Cetus before being saved by Perseus before Andromeda was the Greek hero's wife. Andromeda was the daughter of Cepheus and Cassiopeia. In addition to these names, the constellation is also known in Andromeda mythology as the Chained Maiden, Cepheids, which translates to "daughter of Cepheus," and Persea, which translates to "wife of Perseus." Andromeda's neighbors - Cassiopeia, Cepheus, Cetus, and Perseus - are all connected to the legend of Andromeda in one way or another.

The Positions of the Stars Have Been Plotted and are Visible Today

The positions of the stars have been plotted and are visible today. However, the stars that make up each constellation might be far from one another. Some stars will glow more brightly if they are larger or closer to Earth. In addition, what constellations are visible depends on the time of year.

The Constellation Hydra Occupies 3.16% of the Sky, Making It the Most Extensive in Terms of Area

Hydra's area is 1,303 square degrees, equivalent to 3% of the celestial sphere. It is not hard to believe when one considers the linear expanse of this beast, which spans 95 degrees of the sky and takes nearly seven hours to fully rise into view. On the other hand, the fact that Virgo is the second-largest constellation in terms of area, with a total of 1,294 square degrees, is not as readily apparent. The primary reason for this is that to view the official borders of the IAU. First, you must check a star atlas such as Norton's. Most typical star charts, which only contain the most prominent stars visible to the human eye, do not accurately represent the size of most constellations.

The One That Receives the Most Attention Out of All Eighty-Eight Constellations is the Crux

The Crux constellation, sometimes known as the Southern Cross, is recognized as one of the most prominent in the southern sky. This is the smallest of the 88 constellations; the others are much larger.

The Crux is one of the most well-known constellations in the southern hemisphere, despite being on the smaller side of the constellations. It is easy to recognize because of an asterism that takes the form of a cross known as the Southern Cross. The five stars form this asterism in the constellation that shines the brightest. The constellation is associated with several different narratives and plays an integral part in several different mythologies originating southern hemisphere. The fact that it may be observed at any time of the year and can be located worldwide gives it special significance in Australia and New Zealand.

The Crux is circumpolar in the southern hemisphere, which means it never goes below the horizon. Therefore, it is not visible in the northern hemisphere north of the latitude of +20 degrees, nor in the northern hemisphere north of the freedom of +34 degrees.

Asterisms Include the Big and Little Dipper

Asterisms occur when a tiny star pattern is discovered within a constellation. It includes the Big and Little Dipper.

Hercules, Hydra, Virgo, and Ursa Major are the Five Most Prominent Constellations Seen In the Night Sky

Hercules, Hydra, Virgo, and Ursa Major are the five most prominent constellations seen in the night sky. Cetus is the smallest. Ursa Major, Hercules, Pegasus, Draco, and Leo are the five most extraordinary constellations in the northern hemisphere. The most prominent constellations in the southern hemisphere are Hydra, Virgo, Cetus, Eridanus, and Centaurus. Ptolemy, writing in the second century AD, included all of these Greek constellations in his Almagest. He compiled the list. They have been known ever since the dawn of history. The Lynx is the largest of the constellations that have not been recognized since antiquity.

Aquarius is a Dim Constellation

Although it is one of the largest, most well-known, and oldest recognized constellations, Aquarius is a dim constellation that can be difficult to locate or discern. Aquarius was the personification of Ganymede in Greek mythology, a highly gorgeous young man. Ganymede was invited to Mount Olympus by Zeus because of his excellent looks, and he ended up serving as the cupbearer for the gods there. As a reward for his service, he was given the gift of endless youth and a spot in the starry night sky.

You can see that Aquarius doesn't have any distinguishing traits and doesn't contain any bright stars, although it is both a prominent constellation and a vast one. The line to the right represents Aquarius's right arm, while the expansive form that extends downward represents the water dripping out of the vase and Aquarius's right leg. What is depicted in the image to the right is the portion of the constellation that may most frequently be observed in the night sky. You won't be able to spot the cupbearer in the middle of the city; you'll need to go somewhere with a night sky.

Aquila Can Be Found in the Band of the Milky Way

In Greek mythology, Aquila was the name of the eagle responsible for carrying Ganymede, also known as Aquarius, to the summit of Mount Olympus. In addition, Zeus relied on the eagle to transport his lightning bolts. This constellation may be found in the band of the Milky Way, and its most conspicuous star is Altair, one of the stars that can be seen with the naked eye from the earth. Altair is roughly at the point of the shallow inverted "V" formed by Aquila's upper portion. This is supposed to symbolize the eagle's head and its wings. A line extends downward from Altair to create the eagle's body.

Aries is One of the Twelve That Makes Up the Zodiac

Aries has always been depicted as a ram in mythology, but the myths associated with many other constellations have undergone several variations. This particular constellation is one of the twelve that come together to make up the zodiac, which are the astronomical figures that follow the sun's movement across the sky (known in scientific terms as the ecliptic). The ancient peoples attached significant meaning to the various constellations of the zodiac because of this factor.

Aries is the name of the ram in Greek mythology whose fleece is said to have been transformed into Golden Fleece.

The Canis Major Marks the Beginning of the "Dog Days"

The famous Greek dog Laelaps is symbolized by the constellation Canis Major. Several accounts of how he became a god include the common element that Zeus was impressed by how swiftly he moved. Laelaps is also thought of as one of Orion's hunting dogs, and it is believed to be following in Orion's footsteps across the night sky in their quest for Taurus the bull.

The constellation Canis Major is remarkable because it contains Sirius, the brightest star visible in the night sky. According to tradition, the first sighting of Canis Major in the dawn sky occurs at the end of the summer. This event marks the beginning of the "dog days" of summer. It almost resembles a stick figure when viewed against the backdrop of the night sky, with the brilliant star Sirius serving as its head and the brilliant star Adhara serving as its tail.

Cygnus is Placed Next To His Lyre

In Greek mythology, the form of the swan is assumed by more than one character at a time. For example, at one point, Zeus took the form of a swan to woo Leda, who was later revealed to be the mother of Gemini and Helen of Troy. According to a different legend, Orpheus was put to death, transformed into a swan, and placed in the sky next to his lyre (the constellation Lyra). There is some speculation that the legend of Phaethon and Cycnus contributed to the naming of the constellation.

Chapter 4: Facts About the Solar System

"It's time for the human race to enter the solar system." **- Dan Quayle.**

Sun is the Center of the Solar System

Our sun is the center of our solar system, including everything else in the universe that revolves around it, including the eight planets. However, with all their diversity and intrigue, the primary planets are just the beginning. Earth's neighbors in space include comets, asteroids, dwarf planets, mystery moons, and a whole host of other bizarre occurrences that are so out of this world that they defy explanation.

The Sun is the Most Crucial Source of Light for All the Planets in the Solar System

The sun is the most crucial source of light for all of the planets in the solar system. It is also the object in the solar system that is the most enormous and most massive. The warmth of the sun's rays is something we experience daily; without it, there would be no life on Earth. It is interesting to note that the light coming from the sun does not instantly reach our planet. Instead, because the speed of light is a fixed quantity and the distance between the sun and Earth is approximately 93 million miles (150 million kilometers), it takes the light from the sun about 8 minutes to reach Earth.

Spacecrafts Have the Potential To Transport Contaminants to Inhabitable (Or Already Inhabited) Locations within the Solar System

To this day, researchers have not unearthed any evidence to suggest that there is life anywhere else in the solar system. However, as our knowledge of "extreme" bacteria grows, we better understand how they can survive in environments such as subsea volcanic vents or situations of severe cold, which opens up more possibilities for where they could reside in other worlds.

It is now believed that there is the potential for microbial life on Mars, which has led scientists to take further precautions to sterilize spacecraft before sending them to the planet. However, instead of taking the chance of contaminating Europa's potentially habitable oceans, NASA decided to intentionally crash the Galileo spacecraft into Jupiter.

Mercury Continues To Get Denser

Mercury is currently the dense planet in the solar system after Earth, despite being the minor planet in the solar system overall, excluding the dwarf planet Pluto. And it's only going to get more compact and denser over time.

For a considerable time, scientists were under the impression that Earth was the only planet in the solar system with active plate tectonics. However, this was no longer the case after the Mercury Surface, Space Environment, Geochemistry, and Ranging spacecraft completed its first orbital mission around Mercury. During this mission, the spacecraft mapped the entire planet in high definition and examined the surface features of Mercury.

In 2016, data from the spacecraft revealed cliff-like landforms referred to as fault scarps. Because the fault scarps are relatively small, scientists are confident that they were not formed that long ago, and that the planet is still contracting 4.5 billion years after the solar system's formation. Again, this is because the fault scarps are relatively small.

We Were Wrong in Our Assumptions Regarding the Prevalence of Rings

Since the 1600s, when the first telescopes were created, astronomers have been aware of Saturn's rings. However, the development of spacecraft and telescopes with more excellent magnification capabilities has occurred in the past 500 years. We now understand that every planet in the solar system's outer reaches, including Jupiter, Saturn, Uranus, and Neptune, possesses unique rings.

However, the crews of each world are distinct from one another. Only Saturn's magnificent halo, partially composed of sparkling, reflective water ice, is seen elsewhere in the solar system. On the other hand, the rings of the other giants are almost certainly composed of dust and stony particles.

At the Outer Radius of the Solar System, There Could Be a Gigantic Planet

In January 2015, astronomers Konstantin Batygin and Mike Brown of the California Institute of Technology reported, based on mathematical calculations and simulations, that there may be a giant planet lurking far beyond Neptune. Multiple groups are currently hunting for this hypothetical "Planet Nine," Recent findings imply that discovery may occur within the next ten years.

If this massive object does exist, the motions of some frozen things in the Kuiper Belt, a collection of objects located beyond the orbit of Neptune, would be easier to understand if it is present. If it does exist, the Kuiper Belt is a collection of objects located beyond the orbit of Neptune. Brown has already found several massive entities in that vicinity, some of which have a size equivalent to or even greater than that of Pluto and others that have a size comparable to or smaller than that of Eris.

But there is another possibility that researchers are looking into, which suggests that "Planet Nine" might be a grapefruit-sized black hole that warps space in a manner comparable to how a massive planet might. And yet another team hypothesizes that the strange motions of the distant inhabitants of the Kuiper Belt could be the result of the combined influence of multiple small objects rather than an unknown planet or a black hole in any way, shape, or form.

Neptune is Too Hot

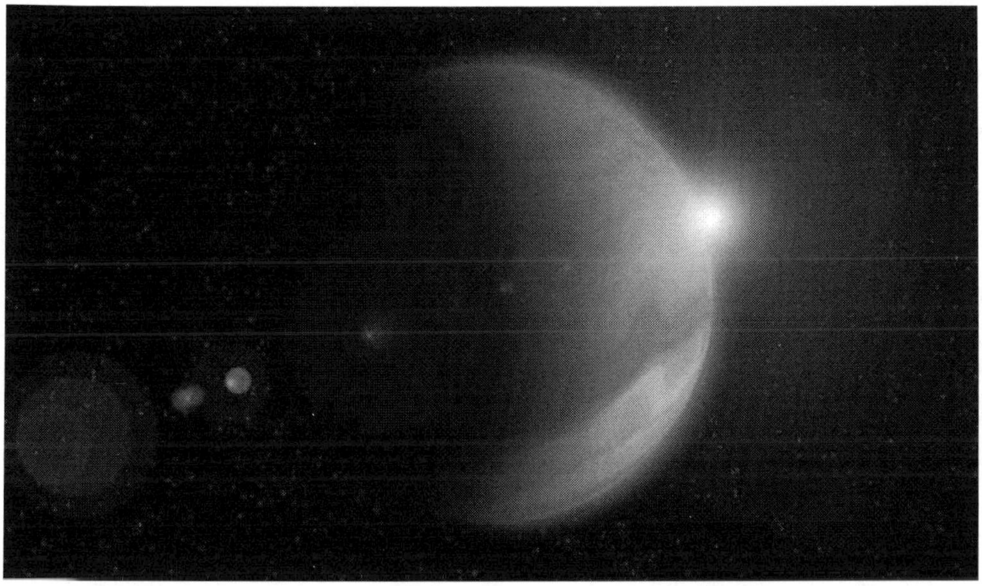

Since Neptune is almost thirty times further from the sun than Earth is, it receives a similarly smaller amount of light and heat. However, it emits much more heat than it can absorb, and there is much more activity in its atmosphere than planetary scientists would predict, particularly in comparison to Uranus, located nearby. It is a mystery to astronomers why Uranus, located far closer to the sun than Neptune, emits almost the same amount of heat as the last planet.

On Neptune, the winds can reach up to 1,500 miles per hour (2,400 kilometers per hour). So, where does all of that energy come from? The sun? The center of the planet? Or is it simply the result of gravitational contraction? Researchers are now investigating the question.

Earth's Van Allen Belts are More Bizarre Than Expected

The multiple bands of magnetically bound, highly energetic charged particles surrounding the Earth are called the Van Allen belts. These belts were named after the physicist responsible for their discovery (named after the discoverer of the phenomenon.) Van Allen Probes, launched in 2012, has given us the most in-depth perspective of the Van Allen Belts we have ever had, Even though we have known about the belts since the beginning of the space age. In addition, they made many unanticipated discoveries when they journeyed along the path.

It is common knowledge that the length of the belts changes in reaction to the sun's activity. At other times, the belts will swell into a single massive entity, while at other times, they will be easy to identify from one another. This phenomenon occurs at both ends of the spectrum. In 2013, researchers announced the discovery of a third radiation belt in addition to the two previously detected. Scientists can increase their abilities to forecast space weather and solar storms if they have a deeper understanding of these belts.

What Came To Pass with Miranda?

Miranda, a dark moon of Uranus that was sighted just once when Voyager 2 got a glimpse of it in 1986, is one of the most peculiar moons in the outer solar system. Miranda was discovered by Voyager 2. On the surface of Miranda, several significant ridges, craters, and other types of disruptions are typically the result of volcanic activity. The cover may have been caused by tectonic action, although Miranda is much too small to create heat at that level.

The researchers believe that the gravitational pull from Uranus could have created the push-pull action necessary to heat, churn, and deform Miranda's surface. To get the answers, however, we will have to launch a second probe into orbit to investigate the uncharted territory in the moon's polar regions.

Saturn Has A Dark Hemisphere

A moon orbiting Saturn Iapetus is distinguished by having one hemisphere that is exceptionally dark and constantly faces away from Saturn and another that is extraordinarily light and always faces toward Saturn. The surfaces of most asteroids, moons, and planets are relatively consistent across their entirety; however, Iapetus occasionally shines brightly enough to be spotted by Giovanni Cassini's telescope in the 1600s, and then it dims down by several magnitudes when oriented in the other direction. Giovanni Cassini discovered this phenomenon.

The most recent scientific findings point to the possibility that Iapetus, commonly known as Saturn VIII, is predominately composed of water ice. Scientists hypothesize that since the moon's night side always faces the light, any water ice that may have been present there melted away, leaving the darker rock in its place. Because dark material warms up more quickly than brilliant, reflective ice, this might have resulted in a positive feedback loop. As the darker, warmer side of the moon lost its ice, it became simpler to heat up each time it faced the sun, which sped up the process of losing more ice.

Titan Has a Cycle of Liquid, But It's Not Water

Titan is another strange moon in Saturn's system. It has a liquid "cycle" that moves things between the surface and the atmosphere. That sounds like how water moves on Earth, but the vast lakes on Titan are filled with methane and ethane, possibly on top of a layer of water.

Researchers hope to use information from the international Cassini mission to figure out some of Titan's secrets and then design a submarine that could one day explore the moon's depths.

Organic Molecules are Everywhere

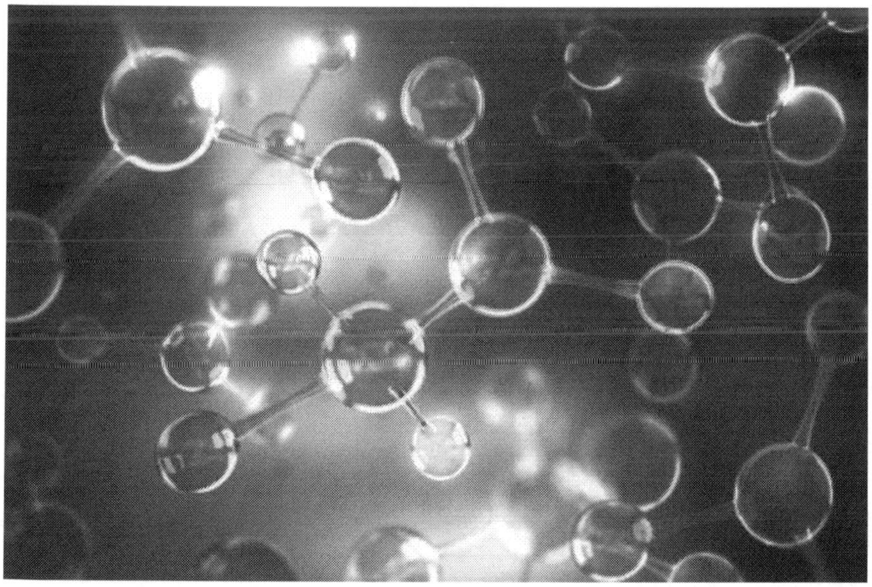

Organics are complex molecules of carbon found in living things but can also be made by non-living processes. Even though organic molecules are common on Earth, they are also found in many unexpected places in the rest of the solar system. On the surface of Comet 67P, for example, scientists have found organic materials. The discovery made it more likely that organic molecules that could have started life on Earth could have come from space.

Organics have also been found on the surface of Mercury, Saturn's moon Titan (which gives Titan its orange color), and Mars.

The Sun's Atmosphere is a Lot Hotter Than the Surface of the Sun

The sun's surface, called the photosphere, is 10,000 degrees Fahrenheit (5,500 degrees Celsius), but its upper atmosphere is millions of degrees Fahrenheit hotter. I wonder why there is a significant temperature difference.

NASA has several spacecraft looking at the sun, and they think they know where the heat is coming from. One of these is a "heat bomb," which happens when magnetic fields in the corona cross and then line up again. Another way is for plasma waves to move from the sun's surface into the corona.

With the Parker Solar Probe, which recently became the first thing made by humans to touch the sun, we are getting closer than ever to understanding what is going on at the center of our solar system.

Venus Shines Brighter Than Anything Else in the Sky

Venus shines brighter than most other things in the night sky. Venus is so brilliant because it is the closest planet to Earth and has thick clouds that reflect about 70% of the sunlight that reaches it. During the short hours after sunset or before sunrise, Venus is frequently the brightest object visible in the sky (other than the moon). It has the appearance of a brilliant star. The planet Venus has the most shining light of any of the other planets.

Chapter 5: Facts About Comets

"I think every time we send a spacecraft to an asteroid or comet, we learn more." - **Carrie Nugent**

Comets Brought Water and Life To Earth

In 2009, a NASA space probe took a sample from a comet called "Wild-2." In that sample, the amino acid "glycine" was found. This amino acid, an essential part of all living things, shows that comets can carry some life.

A recent study suggested that several comet collisions may have brought about 22 trillion pounds of organic material to Earth, along with the energy needed to make more complex molecules from these simple ones. Again, in 2011, a group of scientists found that the water inside a comet has the same chemical makeup as the water in Earth's oceans. This suggests that comets may have brought water to Earth billions of years ago.

Comets Have Their Atmospheres

Comets have their atmospheres because the radiant heat from the sun melts the ice in their nuclei, which then turns into a gas when the comet gets closer to the sun. The gas ejects itself outward from the heart, generating a thin, brilliant atmosphere that can expand to a diameter of 60,000 miles or more. This condition is referred to as a "coma." In 2007, it was discovered that Comet Holmes had an astounding coma with a diameter of around 869,900 miles. A distance that is considerably greater than the diameter of the sun.

Comets Can Strike Our Planet

The chance of an impact depends on how big the object is: the more significant the comet or asteroid, the smaller the opportunity since there are many more small objects in space than big ones. Every day, tons of debris, most of which is smaller than grains of sand, hit the Earth's atmosphere and burn up. These are the "shooting stars" you might see at night. Some larger rocks survive the fiery fall to Earth, and you can see some of these "meteorites" on display in museums. The hazardous objects, which are big enough to cause a regional or global disaster if they hit, might only appear once every few hundred thousand years. Therefore, the likelihood that an object like this will strike Earth in any given year is approximately one in 300,000, which is not a cause for alarm and should not keep you awake at night.

Many Comets Exist and Orbit the Sun

Comets are snowballs made of frozen gas, rocks, and dust that orbit the Sun. When they're frozen, they are as big as a small town. When a comet's orbit brings it near the Sun, it heats up and spews out dust and gases to make a giant glowing head more significant than most planets. The dust and gases form the tail millions of miles away from the Sun. Most likely, billions of comets circle the Sun in the Kuiper Belt and the even farther away Oort Cloud.

The Inner Part of a Comet Looks Like a Dirty Snowball

The solid part in the middle of a comet is called the nucleus. Comet nuclei are a mixture of rock, dust, water, ice, frozen carbon dioxide, carbon monoxide, methane, and ammonia. As a result, they are often called "dirty snowballs," after Fred Whipple's example. "Icy dirtballs" has been a nickname for comets with a lot of dust. The term "icy dirtballs" came about after the July 2005 collision of Comet 9P/Tempel 1 with a NASA Deep Impact mission "impactor" probe. Research in 2014 suggests that comets arc like "deep-fried ice cream" because their surfaces are made of dense crystalline ice mixed with organic compounds, while the ice on the inside is colder and less dense.

All Comets Go Around the Sun

Comets can be of different sizes and shapes, but they all go around the sun. The Kuiper belt is where comets come from. That takes a short time to complete their orbit around the sun. The Kuiper belt is a group of icy objects in the shape of a disc that is beyond the orbit of Neptune. The whole process takes less than 200 years.

The Oort cloud is where long-period comets like ISON come from. The Oort cloud is a group of objects hundreds or even thousands of times farther away from the sun than the Kuiper belt. Comets from the Oort cloud can take millions of years to go around the sun.

Most Comets Have Two Tails

One is made of ionized gas called plasma, and another is made of small solid particles called dust. The tails of comets face away from the Sun. It is because the coma grows into the comet's tail. So, no matter how a comet moves, its tail will always point away from the Sun. Comets have two seats because the Sun affects the gas and dust that escape in different ways, so the tails point in different directions.

The ultraviolet photons from the Sun ionize the gases that are leaving the comet. Once they are ionized, the solar wind carries them away from the Sun in a straight line. It is what makes the plasma tail.

People Have Been Watching Comets for Thousands of Years

Around 500 B.C., some Greek philosophers came up with the word "komotes," which means "long-haired." So, when they saw comets in the sky, they called them "komotes." Halley's Comet is, of course, the most well-known Comet that people have ever seen. Every 76 years, its path brings it close to Earth.

Astronomers in China were said to have written it down for the first time in 239 B.C. But some people say it was seen in Greece in about 467 B.C. Edmund Halley was the first to realize in 1705 that the three comets seen in 1531, 1607, and 1682 were likely the same. So, Halley's Comet will show up in July 2061, which is cool.

People Have Believed Some Crazy Things About Comets

People have believed some crazy things about comets. Comets have been seen as warnings of bad things or signs of good things to come over the years. The Roman emperor Nero thought that a comet meant he would be killed, so he had all his living successors killed. Pope Callixtus III tried to kick Halley's comet out of the church because he thought it was an agent of the devil. On the other hand, William the Conquerer considered a comet a good omen before he invaded England in 1066.

Comet Debris Causes Meteor Showers

Most meteor showers are known or thought to be caused by active or dead comets. However, they are caused by the Earth passing through the orbits of these comets and colliding with their trails of dust and rocks, usually the size of sand grains to pebbles. The showers happen every year, but because the number of meteoroids in the streams (often called "meteor streams") is not always the same, the storms can be quite different from one year to the next.

Comets Can Disappear or Blow Up

A comet less than 100 meters wide will probably blow up high in the atmosphere and not harm. However, if the comet is at least 10 kilometers wide, or if it hits with more than 100 million megatons of force, the damage to the environment around the world will be so bad that it will cause a mass extinction, in which most life forms die. This happened at the end of the Cretaceous period, 65 million years ago, when the dinosaurs died out. The effects of a significant impact on the environment come in many forms. The worst products are a global firestorm caused by falling impact debris and a global haze caused by dust in the stratosphere that lasts for many months.

Comets Have a Very Elliptical Path Around the Sun

Comets have a very elliptical path around the Sun. As a result, they can spend hundreds of thousands of years deep in the solar system before their perihelion when they circle back around the Sun. Comets move according to Kepler's Laws, which say that the closer they are to the Sun, the faster they move. Even though a comet is far from the Sun, it looks like a dirty snowball that is several kilometers across. But as it gets closer to the Sun, its surface heats up, causing its materials to melt and turn into vapor. This is what makes the comet's tail. The spacing between the Earth and the Sun is the same length as the tail of a comet.

Four Parts of a Comet Can Be Seen

Four parts of a comet can be seen: the nucleus, the coma, the ion tail, and the dust tail. The nucleus is a solid body usually a few kilometers across and made up of a mixture of volatile ices (mainly water ice) and silicate and organic dust particles. The coma is the free-flowing atmosphere around the nucleus. It forms when the comet gets close to the Sun and the volatile ices sublimate, carrying dust particles mixed with the ice in the middle. The dust tail is made up of these dust particles, which are pushed back by the pressure of the Sun's radiation to form a long, curved tail that is usually white or yellow. Finally, the volatile gases in the coma make up the ion tail when they are ionized by the Sun's ultraviolet photons and blown away by the solar wind. Ion tails almost always point away from the Sun and glow blue because CO^+ ions are present.

When a Comet Gets Close to the Sun, It Makes a Halo

As comets get closer to the Sun, the heat from the Sun starts to melt the ice, forming a coma, a fuzzy, bright area of vaporized gas around the comet's core. A hydrogen halo is a layer of hydrogen gas that goes around the coma and can be up to 1010 meters across.

There are Two Kinds of Comets

Comets spend most of their time in the far reaches of the solar system, far from the Sun. The Kuiper Belt and the Oort Cloud are where most of them come from. The Kuiper Belt is a disk of primarily icy bodies that goes from about the orbit of Neptune, which is, on average, about 30 AU to about 50 AU from the Sun. The Oort Cloud is at the edge of the Sun's gravitational pull (about 50,000 to 200,000 AU). It is made up of icy bodies and is split into two parts: the inner, disc-shaped Hills cloud and the outer, spherical cloud.

Cloud Oort

The Oort cloud has never been seen, but scientists think it is a sphere of icy objects like comets that orbit the Sun between 3000 and 100,000 AU away. It is also believed to be where many of the comets with long orbital periods came from. The objects in the Oort cloud probably started closer to the Sun, around Uranus and Neptune, and were then pushed out to where they are now by the planets' gravity.

Astronomers think the Oort cloud comprises about 1012 to 1013 objects with a combined mass of about 100 Earth masses. Things in the Kuiper belt are affected by how the planets pull on them. Between 50 and 2000, AU is a part of the Oort cloud, where things are not affected by the world. However, objects in the cloud are affected by galactic tides from 2,000 to 15,000 AU, and from 15,000 to 100,000 AU in the outer Oort cloud, objects are affected by the gravity of other stars. Outside the Oort cloud, the Sun's gravity is not strong enough to keep things in orbit.

The Kuiper is a Large Area of Space Past Neptune

The Oort Cloud, the heliosphere, and Jupiter's magnetosphere are the other three most significant structures in our solar system. It looks like a round, puffed-up disk or a donut. Its inner edge starts at about 30 AU from the Sun, where the orbit of Neptune starts. The distance from Earth to the Sun is measured in astronomical units (1 AU). Around 50 AU from the Sun is where the central inner part of the Kuiper belt ends. A second area called the "scattered disk" overlaps the outer edge of the central part of the Kuiper Belt. It reaches nearly 1,000 AU, and some bodies are on orbits that go even further.

A Lot of Things in the Kuiper Belt Have Moons

Many KBOs have moons, which are much smaller bodies that orbit them or are made up of two objects. Binaries are two objects of the same size or mass and orbit around a point between them. This point is called their shared center of mass. Some binary numbers touch, making a shape like a peanut. This is called a "contact binary." Kuiper Belt objects with moons include Pluto, Eris, Haumea, and Quaoar.

Kuiper Didn't Find the Kuiper Belt

The Kuiper Belt is named after the astronomer Gerard Kuiper, who wrote a paper about objects beyond Pluto in 1951. Kuiper's work didn't predict the number of things we see in the area named after him or, more importantly, how they relate to Neptune. The belt's inner edge is set by Neptune's orbit, not Pluto's, and most of the belt's shape is due to Neptune's gravity. But astronomers knew about Kuiper and his ideas, so the general idea of the belt came to be credited to him.

A Comet's Impact with Jupiter

In 1994, a comet called Shoemaker-Levy 9 spectacularly crashed into Jupiter. The planet's pull on the comet tore it apart because of its mass. This collision was seen to have about 21 visible impacts.

The enormous collision ever seen led to a massive fireball that rose about 1,800 miles into the air. The crash made a vast dark spot up to 12,000 kilometers across. Scientists say the force of the collision was about the same as 6,000 gigatons of Trinitrotoluene (TNT).

Comet Hale–Alan Hale and Thomas Bopp Made Bopp on July 23, 1995

Comet Hale–Bopp was made by Alan Hale and Thomas Bopp on July 23, 1995, before it was big enough to see with the naked eye. It's hard to tell how bright a new comet will get, but Hale–Bopp was brighter and more colorful than most people thought it would be when it reached a magnitude of 1.8 on April 1, 1997, when it passed perihelion. It was so big that people could see it with their own eyes for 18 months straight. The old record holder, the Great Comet of 1811, was half as long as this. So, people started calling Hale–Bopp the "great comet" of 1997.

Kuiper Belt Shares Similarities with the Central Asteroid Belt

Astronomers think the icy objects of the Kuiper Belt are remnants from the solar system's formation. Similar to the relationship between the central asteroid belt and Jupiter, it's a region of objects that might have come together to form a planet had Neptune not been there. Instead, Neptune's gravity stirred up this region of space so much that the small, icy objects couldn't coalesce into a giant planet.

The Amount of Matter in the Kuiper Belt May Be Only a Tiny Part of What It Used to Be

The amount of matter in the Kuiper Belt may be only a tiny part of what it used to be. One well-supported theory called the "Nice Model," after the city of Nice, France, says that the shifting orbits of Jupiter, Saturn, Uranus, and Neptune could have taken away most of the original material, which was probably 7 to 10 times the mass of Earth. The Kuiper Belt is slowly breaking apart right now. Objects there sometimes crash into each other, creating smaller KBOs, some of which may turn into comets and dust carried away by the solar wind. The mass of everything in the Kuiper Belt is thought to be no more than 10 percent of the group of Earth.

Kuiper Belt is a Source of Comets

As it slowly wears away, the Kuiper Belt is a source of comets. When Neptune's gravity can push two KBOs to crash into each other, the pieces are made into orbits that send them toward the sun, whereas Jupiter's gravity keeps them in short loops that last 20 years or less. These are called Jupiter-family comets with short orbital periods. Most comets lose their volatile ice quickly because they often come close to the sun. This means that they eventually become dormant, or dead, comets that don't move much. Scientists have found that some asteroids close to Earth are burned-out comets, most of which would have come from the Kuiper Belt. The Oort Cloud is the other place where comets come from. This is where most long-period comets with highly tilted orbits come from.

Astronomers Didn't Know They'd Found Kuiper Belt for a Long Time

Pluto was the first object in the Kuiper Belt to be found. It was in 1930 before astronomers had any reason to think there would be a lot of icy worlds beyond Neptune. At the time, scientists didn't know much about the outer parts of the solar system, so they didn't think Pluto might have a lot of friends. So, even though Pluto had an oddly elliptical and tilted orbit, it made sense to think of it as a separate planet at the time. However, the second KBO wasn't found until 1992, 62 years later. It made it clear that Pluto isn't the only thing out there.

Chapter 6: Facts About the Moon

"The Moon will guide you through the night with her brightness, but she will always dwell in the darkness to be seen." - **Shannon L. Alder**

The Moon Travels Around the Earth in 27.3 Days

In about 27 days, the Moon makes one trip around the Earth. The sidereal month is the name for this period. In 29.5 days, the Moon goes through all of its phases. A synodic month is the amount of time it takes for all the cycles to happen. The Moon moves at an average speed of 2,288 miles per hour, but at different points in its orbit, it moves at different rates. It moves faster when it's close to Earth but slower when it's far away. During one orbit, the Moon goes about 1,423,000 miles around the Earth.

The Sun Dwarfs the Moon in Size

When viewed from above, the Sun and Moon appear roughly the same size. This is because the Sun is 400 times farther away and 400 times more potent than the Moon. Another interesting fact is that the distance between the Sun and the Moon is roughly twice as long as the distance between the Sun and the Moon. If the Sun were to be placed where the Earth is, the Moon would be a little over halfway to the Sun's surface. The Sun is also so massive that light takes nearly four and a half seconds to traverse across its surface. I'm not sure if that tells more about the size of the Sun, the speed of light, or both.

When a Rock Hit Earth, the Moon was Made

Scientists think the Moon was made when a large object hit the Earth 60–175 million years after the solar system's birth. They can get this estimate by looking at rocks on Earth.

As large planetesimals grow, they get hot enough to melt because of repeated impacts and the radioactive decay of elements inside their minerals.

The Moon is in Charge of the Tides

The Moon and Earth pull on each other because of their mass. The Moon's pull on Earth makes the oceans bulge on the side closest to the Moon and the side farthest from it. These bumps cause high tides. Low tides happen at the low points.

Earth's Moon is the Only Place Other Than Earth Where People Have Walked

So far, humans have only set foot on the Moon. People call Earth's only natural satellite "the Moon" because they didn't know there were other moons until Galileo Galilei found four moons orbiting Jupiter in 1610.

On the Moon, You Can't Breathe

The Moon has something called an exosphere, which is a thin layer of air. Astronauts need space suits and helmets that let them breathe because the air in space is not breathable.

The Moon is Moving Away from Us

The Moon used to be about ten times closer to Earth than it is now and still moves away from Earth by 38cm every year. That's about how fast your fingernails grow.

Moonrise

The phase of the moon when it is in line with the sun and its dark side is facing the earth.

The Growing Moon

A waxing crescent moon means that the moon is getting brighter, but it's still less than half lit up.

The First Quarter

The first quarter is one of the moon's four main phases. It happens between the new and full moon when half of the moon's lit surface can be seen.

Waxing Gibbous Moon

When it gets dark, there is a moon getting bigger and rounder. It makes the early evening bright. It looks more radiant than half but less bright than complete. Waxing gibbous moons are between the first quarter and full moons.

Full Moon

When its whole disk is lit up, the moon happens when it is opposite the sun.

Waning Gibbous

After the Full Moon, this phase lasts until the Third Quarter, when half of the Moon's face is still lit up.

Chapter 7: Facts About Meteors and Other Sky Shows

"Great men are meteors, consuming themselves to light the world." **- Napoleon Bonaparte**

Scientists Think That Many Meteorites Come from Asteroids

Most meteorites on Earth come from broken-up asteroids, but some come from Mars or the Moon. Small pieces of Mercury or Venus could also have landed on Earth, but none have been found for sure. Scientists can figure out where meteorites come from based on several clues. Photographs of meteorite falls can be used to figure out their orbits and project their paths back to the asteroid belt. They can also compare the way meteorites are made to the way different types of asteroids are made. Finally, they can also figure out how old the meteorites are, up to 4.6 billion years old.

Meteorites are Chunks of Rock that Crash Upon the Surface of the Earth After Traveling Through Space

This is typically a solid chunk of debris left behind by extraterrestrial objects like comets, asteroids, or even meteoroids. These are regarded as meteorites if they have survived the journey from outer space through an atmosphere and onto the surface of a planet or moon without being broken apart.

Because of the Crater It Made, the Barringer Meteorite is One of the Most Well-Known Meteorites

The Barringer meteor impact crater in Arizona is one of the most famous and visited impact craters in the United States. Scientists almost all agreed it was the first structure caused by a meteorite. The cavity is 0.75 miles (1.2 km) wide and was formed 49,000 years ago. However, it was challenging for people to accept it as an impact crater. Daniel Barringer, the head of a mining company, was the most vocal supporter of the meteorite impact theory. He argued for years against Grove K. Gilbert, the influential leader of the U.S. Geological Survey, who said that the crater was a volcanic feature.

Daniel Barringer lost the argument for many years because, according to his model, the hole should have been covered by a sizeable iron-nickel deposit from the meteorite. However, this deposit was never found because the impact's heat and energy vaporized the meteorite. By 1930, however, there was enough other evidence that the crater was caused by a blow from space to convince the scientific community of this.

Meteoroids, Meteorites, and Meteors Have Different Names That Show Where They are Now

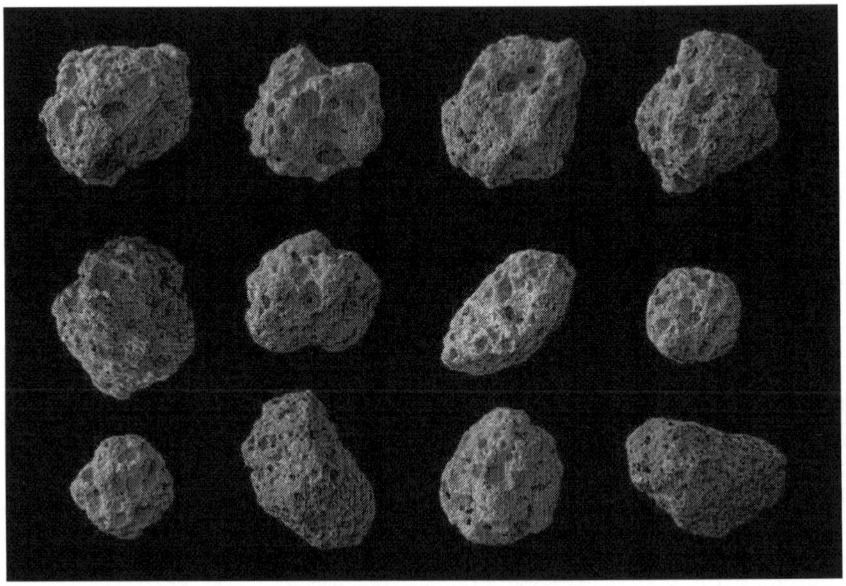

They all have to do with the bright flashes of light called "shooting stars" that sometimes move across the sky. But we have different names for the same thing based on where it is.

Meteoroids are small space objects, like dust grains or minor asteroids. Imagine them as "rocks from space."

Meteors are fireballs or "shooting stars" made by meteoroids that fly into Earth's atmosphere (or the atmosphere of another planet, like Mars) at high speed and burn up. The meteorite is the name for a meteoroid that makes it through the air and lands on Earth.

Meteor Showers Show How Beauty Can Come from Ashes

Lang says that meteor showers show how beauty can come from ashes. "Meteor showers are like the spiritual insights and wisdom we gain when we look deep into our own and the world's shadows. They also ask us to think of and believe in new things. They are like short bursts of "aha!"," she says.

At the end of July, there were two meteor showers (double rainbow, but make it space). The rain came from the Aquarius and Capricorn constellations. Lang says that Aquarius is the sign of community and new ideas, while Capricorn is the sign of "authority, structure, institutions, ambition, and our public life." So even though the rain could be seen last night and early this morning, it started in early July and will continue through August.

When a Meteoroid Has a Diameter of More Than 10 Meters, It is Called an Asteroid

Asteroids are rock-like objects that circle the Sun. But because they are so small, they can't be called planets.

Millions of asteroids circle the Sun, and even though they are small, they can be dangerous if they hit Earth. The weight of all asteroids together is less than that of the moon.

The Majority of Scientists Subscribe to the Theory that a Massive Meteor Impact Was the Cause of The Extinction of Dinosaurs

The Chicxulub crater is the primary evidence scientists utilize to back up this theory. According to the details of the collision, it had a diameter of 11-81 kilometers and created an energy explosion equivalent to up to 21-921 billion Hiroshima bombs.

It is believed that this collision was responsible for the extinction of all living things that existed throughout the prehistoric period, including plants and dinosaurs.

Andrew Ellicott Watched the Leonids Meteor Shower from a Ship Anchored Off the Coast of the Florida Keys

An early American astronomer named Andrew Ellicott watched the Leonids meteor shower from a ship off the coast of the Florida Keys. In his notebook, Ellicott recalled thinking that "the whole heaven appeared as if lighted with sky rockets, flying in an infinite number of directions" and that he was "constantly expecting some of them to fall on the vessel." They persisted until the light of the dawn, which appeared just after daybreak, ended it. The item in Ellicott's journal is the first record that is believed to exist of a meteor shower occurring in North America.

The Leonid meteor shower is an annual phenomenon that is considerably augmented approximately once every 33 years by the appearance of the comet Tempel-Tuttle. The Leonids are named after the constellation Leon. When the comet returns, the Leonids have the potential to produce rates of up to several thousand meteors per hour, which can cause the sky on a clear night to become illuminated. Following the return of the comet Tempel-Tuttle in 1833, Ellicott noticed one such manifestation of the Leonids shower. This event is credited with motivating the first organized study of meteor astronomy.

A Total of Approximately 40 Meteorite Craters with a Diameter of More Than 12 Miles Have Been Found By Researchers

In total, around one hundred twenty impact craters have been discovered over the surface of the Earth. The most recent impact documented to have caused a hole of this size was in 1908.

According to the reports, this occurred in a region of western Siberia known as Tunguska, which is entirely devoid of inhabitants. The diameter of the impactor was around 60 meters, and it is made up of several separate parts.

Solid Objects That Penetrate the Atmosphere of the Earth But Have Not Yet Reached Its Surface Are Referred To Be Meteors

A wide range of sizes can be found among meteors. These can be as minute as a particle of dust that we can see when held up to the light or as large as a mansion. The word meteor is derived from the Greek word meteor, which means "falling star."

The literal meaning of this is something that occurs in the sky. The ancient Greeks used this word to describe the bright smears of light that could be seen throughout the night sky as meteors collided with the Earth's atmosphere.

Radio Signals Sent Out by the National Weather Service are Reflected Off the Ionized Trails Left Behind By Meteors

Meteor burst communication and meteor scatter communication are two more names for this type of communication. This is used to collect data from automated weather stations and data from other government entities in the United States on an as-needed basis.

The vast majority of meteors are capable of scattering communications carried out between radio stations actively engaged in a particular schedule of transmission and receiving intervals. When gathering data in remote areas, many other countries and regions within the United States use remarkably similar approaches.

Meteors Can Look Like They Have Different Colors

If you look at a meteor out of the corner of your eye, it might look like a flash of white light, like how stars look. Meteors, on the other hand, can look like they have different colors if the sky is clear or a picture is taken. Meteorologist Dave Samuhel from AccuWeather said, "The color of the light that meteors give off depends on what chemicals they are made of."

Samuel said, "As the meteors burn up as they enter the Earth's atmosphere, the different chemicals in them make them turn different colors."

For example, mostly made of calcium, meteors will look purple or violet, while those mostly made of magnesium will look green or teal. What a meteor is made of is not the only thing that affects what color it looks like. The color can also be changed by how fast the meteor hits the Earth's atmosphere.

The Heat from the Meteor

Meteorites don't look the same on Earth as when they were floating through space. In the 10 to 15 seconds it takes a meteorite to travel through the atmosphere, air friction heats its surface to 1,800 degrees Celsius. This friction can melt the meteorite and take away up to 90% of its original mass, leaving exciting surface features. The surface of a meteorite starts to change and break down as soon as it hits the ground.

Meteorite Drippings

Some meteorites, like Stannern, have tiny bumpy lines running down their sides. These are called "flow lines," which look like wax drips on a candle. Meteorites that stay the same way as they move through the atmosphere get flow lines. Air pressure melts the surface and pushes the droplets backward. As the rock cools, the last drops of water that were flowing freeze into flow lines.

The Region Between Jupiter and Mars in Our Solar System Contains Most of the Known Asteroids

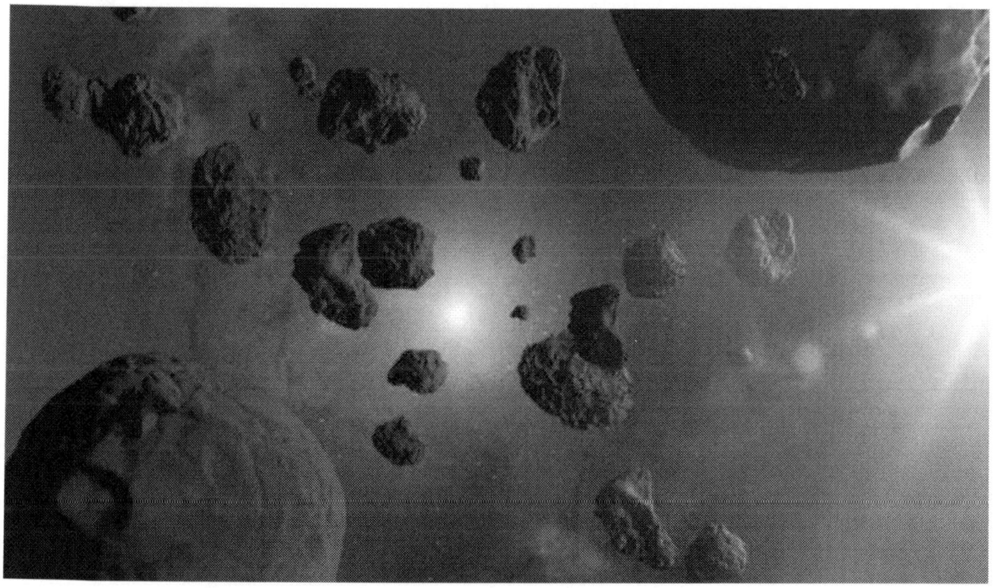

The vast majority of asteroids now known have orbits that take them between Mars and Jupiter, and these orbits are often stable. On the other side, it is thought that there are millions more small asteroids and anywhere from 1.1 million to 1.9 million asteroids that have a diameter of more than 0.9 kilometers (1 kilometer). The asteroid belt contains both types of asteroids. In the early stages of the solar system's history, the gravity of a newly created Jupiter prevented the formation of planetary bodies in this region. Instead, it caused tiny bodies to collide, breaking them into the asteroids we observe in the solar system today.

The Hoba Meteor is the Largest Meteorite Ever Discovered Anywhere on Earth

The farmer had unearthed a meteorite made of iron that weighed 66 tons and was the largest meteorite ever observed. It was also the most significant piece of iron ever discovered close to the surface of the Earth. It looks like a table and is roughly 9 feet in length, 9 feet in width, and 3 feet in depth. The animal was discovered on a "Hoba West" farm where the name "Hoba" originated.

There is a common belief that Hoba crashed into Earth approximately 80,000 years ago. It is composed primarily of iron (about 84%), nickel (16%), and trace amounts of cobalt and other elements. It is possible that the meteorite was much more giant than 66 tons when it landed on Earth but has shrunk significantly due to oxidation, which is indicated by the high concentration of iron oxides in the soil next to the meteorite.

Meteoroids Can Be of Different Sizes, But They are Usually Too Small for People to Notice When They Fall

Most of the time, these meteorites land in bodies of water and are never found or studied. However, research shows up to 300 tons of space dust and small meteorites fall to Earth daily.

Even though many things fall from space daily, it is about 90–95% safe. Most meteors burn up entirely in the atmosphere before hitting the ground.

Meteors are Made When Air Particles and Pressure Rub Against Each Other As They Enter the Earth's Atmosphere

A meteor is a bright streak in the sky that is created by a meteoroid crashing into the atmosphere of Earth. Rock and iron fragments that orbit the Sun are referred to as meteoroids. Most meteoroids are composed of pieces of rock that were created when larger asteroids collided with one another.

However, as they orbit the Sun and shed dust and debris, comets give off particles that eventually become meteoroids. When a meteoroid enters the high atmosphere of Earth, it gets heated up because of the friction generated by the rushing air. The meteoroid will absorb this heat as it travels into space. The formation of a meteor occurs when the temperatures of the gases surrounding a meteoroid rise to a point where they cause the gases to light brilliantly. The end product of this process is the production of a meteor.

When a Meteoroid Has a Diameter of More Than 10 Meters, It is Called an Asteroid

Asteroids are rock-like objects that circle the Sun. But because they are so small, they can't be called planets. Millions of asteroids circle the Sun, and even though they are small, they can be dangerous if they hit Earth. The weight of all asteroids together is less than that of the moon.

A Meteor Shower Might Encourage You To Disengage More From the Real World and Devote More of Your Attention to The World of Ideas and Fantasies

Capricorn is the zodiac sign for people born in January., which is an earth sign. You are not alone. As a result of this meteor shower, you can move more into the domain of ideas and dreams and less into the functional area. Then, you can return to reality and concentrate on how those thoughts can be implemented at the appropriate time. But for the time being, keep an open mind and be willing to investigate everything that is up for you. Additionally, this is an opportunity for education and research. It's possible that your thinking is more receptive than usual.

The Meteorite with the Smallest Size

The population of microscopic particles that fill interplanetary space is the source of micrometeorites. Micrometeorites are a subtype of meteorite that can range from a few hundred micrometers (m) to as small as around 10 m. Therefore, they are considered to be relatively insignificant.

The vast majority of the meteorites discovered on Earth are believed to be bits of asteroids that orbit in the inner region of the central asteroid belt, according to research conducted in the laboratory by astronomers and theorists. This region of the main asteroid belt is situated between approximately 2.1 and 3.3 astronomical units (AU) from the Sun.

The Acquisition of Meteorites

Meteorites are traditionally referred to by the name of a geographical feature connected to the region in which they are discovered. However, a short time ago, concerted efforts were only made to find and retrieve them. It is because meteorites fall more or less evenly across the surface of the Earth, and there was no obvious way to anticipate where they would sink or where they could be found. So instead, the item was taken to a museum or a private collector whenever a meteorite was witnessed falling to the ground, or someone struck an unusual-looking rock by coincidence.

How Old are Meteorites?

Radioactive isotopes, radionuclides, were in the planets and asteroids when they first formed. Radionuclides break down at regular rates. The amount of time needed to complete one-half of a task radionuclide's atoms to disintegrate is referred to as the radionuclide's "half-life." This is a common way to measure how fast it breaks down. Many radionuclides have half-lives that are the same as or longer than the age of the solar system. Because of this, they are often called "long-lived" radionuclides. Because they last so long, they can still be found on Earth and in meteorites, often used to date rocks and meteorites.

The Ingredients and Dimensions of the Meteoroids Determine the Length of the Light Trail They Leave Behind

The meteoroid's journey through the atmosphere produces an illuminated light trail. The meteoroid starts to burn up as it descends, and this process speeds up as it gets lower.

The method in question is known as adiabatic compression. This is because the length of the visible tail can reach up to more than 150 million kilometers. In addition, the tail flows in the opposite direction of where the Sun is at all times.

Chapter 8: Facts About Astronauts

"As I mentioned, it's been a long, boring drive. And I'm still on the outward leg. But hey, I'm an astronaut. Long-ass trips are my business." — **Andy Weir, The Martian**

The Physiological Requirements for Space Flight and How They Can Be Satisfied

In a recent interview with MEGA, Pedro Duque offered his argument that, from a purely physical aspect, extraordinary circumstances are not required for space flight. This was in response to a question about whether or not space travel requires any unusual circumstances. Not only do you need to be in good health, but you also need to demonstrate that you can bear the three gravitational forces of pressure in the heart, as stated by the astronaut.

The First Such Journey into Space by a Human Being

The Russian cosmonaut Yuri Gagarin, who made history as the first person to journey into space on April 12, 1961, aboard the ship Vostok 1, can be considered the story's hero. Gagarin walked around the Earth at a pace of 27,400 kilometers per hour, and the journey took 108 minutes. His words used to describe the encounter indicated the immense surprise that he was feeling: "The planet we live on is blue! When viewed from a cosmic height, the world can be seen in all its detail, with mountains, coasts, and islands easily separated from one another." In his formal report on the flight, he elaborated on this point much later.

NASA MMU Backpack

This propulsion backpack weighing 140 kilograms was worn by astronauts of the space shuttle and is also known by its other name, the Manned Maneuvering Unit. The high-pressure nitrogen jet and the EMU space suit are used to propel it after it has been placed inside the case. It comprises two aluminum tanks, each holding 5.9 kilograms of nitrogen at a pressure of 20.7 kilopascals. This nitrogen is sufficient for extravehicular activities for six hours (EVA).

How Do Astronauts Relieve Themselves?

They utilize these unique diapers known as MAG for journeys that are only a few days longer. In addition, the ships have special toilets that can function appropriately in zero gravity throughout prolonged excursions. These do not remove garbage from the boat by absorbing it via the water; instead, they operate as powerful vacuum cleaners that remove everything floating in the ship's environment.

A Spanish Astronaut Carried Out the First Space Mission

Regarding birthplace and nationality, Pedro Duque was the first person of Spanish descent to travel into space. In October 1998, he was one of seven astronauts who embarked on the STS-95 mission aboard the shuttle Discovery. Their primary objective was to complete the task. To our great fortune, right within that same month, Pedro Duque made the time to offer MEGA Interesting his final interview before he left.

Cleaning Up Outer Space

Because practically all the water astronauts consume in space must be transported from Earth, careful management of this limited and valuable resource is essential for their survival. Because of this, rather than having a shower, they use a damp cloth to lather themselves up.

Fire-Resistant Suits

This suit model is constructed out of synthetic materials resistant to heat, so it does not catch fire or melt when it comes into touch with flames. Proximity suits, worn by firefighters and people who respond to air mishaps, can endure temperatures of up to 260 degrees Celsius. Those used by astronauts must, of course, be capable of protecting them in extreme temperature circumstances, such as the great cold of space (-270 degrees Celsius) and the intense heat of re-entry into the atmosphere of Earth (1,260 degrees Celsius).

Orlan Suit

The Russian spacesuit model in question makes it possible for astronauts to spend more than seven hours away from their base of operations. During this time, they can install new hardware or carry out repairs. 0.4 atmospheres is the pressure that can be found within the suit. The astronauts' use of spacesuits provides several benefits. When doing spacewalks, astronauts are exposed to a wide range of temperatures. In Earth's orbit, temperatures can drop to as low as minus 260 degrees Fahrenheit at their coldest. When exposed to direct sunshine, they can reach temperatures up to 250 degrees. The astronauts are shielded from these severe heat conditions by their spacesuits.

The Risks To One's Health Posed By Space Flight

It is common for astronauts to experience a loss of muscle mass when in space because they do not exert the same level of power required on Earth to overcome the resistance caused by gravity. The lack of seriousness causes the fluids to flow from the lower extremities to the upper half of the body, which might result in their faces becoming puffy at times. In addition, their bone density decreases by more than 1% every month while in space, which is another injury they sustain. After returning from the area, astronauts often have trouble with everyday activities such as walking and completing other routine tasks.

Preparation For Excursions into Space

Did you know that most of the time spent practicing for space travels takes place underwater? The weightlessness that comes with floating is one of the reasons why many professionals believe that floating is the most effective approach to train for tests that take place on Earth, as outlined by NASA.

Valentina Tereshkova

She was the first woman in the history of space travel and was born on March 6, 1937. Her full name was Valentina Vladmirovna Tereshkova. A native of the Soviet Union, Tereshkova made history on June 16, 1973, at 9:23 a.m., when she became the first civilian to fly into space. She did this when she was a passenger on the Vostok 6 spacecraft. She was chosen out of a pool of more than 400 applicants, and throughout her mission in space, which lasted for three days, she performed 48 orbits around the Earth.

Customs in Russia

Russia is the only country that currently deploys armed astronauts into space. The only reason for this tradition is to ensure the protection and safety of the pilots. For example, if the spacecraft needs to make an emergency landing in a desert or the middle of a forest, the astronauts can defend themselves if wild animals attack them.

Expanding Toward Outer Space

The absence of gravity causes a reduction in the amount of pressure exerted on the spine, resulting in a gain of around five centimeters in height for astronauts.

On the surface of the Moon, astronaut Alan Shepard, who was a part of the Apollo 14 mission team, played a round of golf. Unfortunately, his third effort at striking the ball resulted in it being hit so far away that its current whereabouts are uncertain.

There are Several Limitations to Board

Granulated foods like sugar, salt, and pepper are not permitted in the hallways at any time. As an exception to this norm, salt and pepper are typically presented to the table in liquid form.

Alterations Also Occur In the Heart

An astronaut's heart will grow almost 10 percent more spherical after extended periods spent traveling through space. According to those knowledgeable in the subject, this results from being subjected to microgravity for a considerable time. This is problematic since it increases the risk of developing cardiac problems.

The Nails Come Loose

As a result of their heaviness and bulkiness, astronauts' gloves with their space suits directly impact their fingernails. The astronauts squeeze their fingers so tightly that they cut off the circulation to their fingertips, but at the same time, the intense pressure pushes the tips of each of their fingers. This results in the astronauts losing feeling in their fingertips. Because of this, their hands are covered in blisters, and their fingernails gradually start to fall off.

Before embarking on their mission, some astronauts go as far as to remove their fingernails and toenails to reduce the risk of harm and infection.

The Late Bruce McCandless

Some people would describe him as "intrepid" and "brave," while others would describe him as "reckless" and "foolish." The longest spacewalk was completed by Bruce McCandless, born the same year as Tereshkova (in 1984). McCandless moved more than one hundred meters from the ship while maintaining no connection to the vessel. He navigated his environment with the assistance of a nitrogen-powered propulsion system.

What Is the Typical Weight of a Space Suit?

On the ground, a spacesuit typically weighs about 280 pounds, even when it is empty and not worn by an astronaut. A spacesuit does not affect the wearer's weight in space because there is no gravity. The cooling system in an astronaut's suit is controlled by more than 90 meters of pipes that run throughout the suit. The importance of a space suit is around 127 kilograms, making it a highly cumbersome item.

Laika, the Dog Who Lived in Space

Laika, a dog with Soviet roots, is the first terrestrial living being to journey into space. She was known as "the first dog in space." When launching preparations for Sputnik 2 began on November

3, 1957, Laika was already on board. Unfortunately, the animal was subjected to such a high amount of stress that it passed away within a few hours of being sent into space (the period for this is estimated to be five hours). It makes the story much more sorrowful than it initially appears.

Snoring Disappears

The absence of air in space prevents sound waves from traveling and, as a result, from being perceived by humans. In addition, outer space stops the typical snoring that individuals are known to do because there is no medium in which waves can be transported, such as air.

A Fitness Center Exists Amid Space

Because muscles readily atrophy in space, astronauts must maintain their muscular mass through regular exercise. Because of this, there is a fully functional gym on the International Space Station to exercise.

Shattered Nighttime Visions

Trying to get some shut-eye in space presents its unique difficulties. First, getting a "good night's sleep" is difficult as the sun rises and sets nearly every 90 minutes, making it impossible to keep track of the time.

A Total of Twelve Men Have Stepped Foot On the Surface of the Moon, with Two Astronauts from Each of the Six Apollo Missions

The twelve individuals invited to join the most elite club in the annals of human history shared several characteristics.

All of them hailed from very technological families, and all of them, except one, studied either aeronautical or astronautical engineering. Many of them had participated in the Boy Scouts when they were younger, and an even more significant number were active members of the fraternities they belonged to at their respective universities. They all went on to get additional degrees, most of which were earned at military test pilot schools, and almost all saw active duty during the cold war. Many were tasked with flying nuclear missiles behind enemy lines while serving in this capacity.

These high-risk occupations routinely claimed the lives of individuals to the left and right of them, and in many cases, luck was the only thing that kept them alive long enough to apply to NASA.

It is possible that guys who went through a selection procedure designed to find those with "the right stuff" might lead remarkably similar lives. However, in the years following Apollo, it became apparent that no two individuals were cut from the same mold, despite their resumes sharing some commonalities. They were Neil Armstrong, who was the commander of Apollo 11, Buzz Aldrin, who was the lunar module pilot for Apollo 11, Pete Conrad, who was the commander of Apollo 12, Alan Bean, who was the lunar module pilot for Apollo 12, Alan Shepard, who was the commander of Apollo 14, Edgar Mitchell, who was the lunar module pilot for Apollo 14, David Scott, who was the commander of Apollo 15, Jim Irwin, who was the lunar module pilot for Apollo 15, John Young, who was the commander of Apollo 16, Charlie Duke, who was the lunar.

A More Significant Number of Men Than Women

It is not as straightforward as you may believe in understanding why fewer women are working in the field of astronautics than men. Because the maximum safe level of exposure to radiation in space is lower for females than for males, this immediately becomes a concern for the health of female astronauts. This is because females have a higher risk of being exposed to radiation from space. NASA researchers have demonstrated that prolonged exposure to cosmic radiation, whether on the surface of some planets or in deep space, raises the risk of developing diseases such as cancer.. This is true regardless of the location of the radiation exposure. It is the case irrespective of whether the exposure occurs on the planet's surface or in space itself.

There are Six Astronaut Movies That You Should Check Out

If you have a deep interest in the fields of astrophysics and space exploration, several feature films were inspired by this fascinating topic that you shouldn't pass up seeing, even though most of them are pretty old: Interstellar was directed by Christopher Nolan in 2014, Duncan Jones directed Moon in 2009, Apollo 13 was directed by Ron Howard in 1995, Ridley Scott directed the Eighth Passenger in 1979, Andréi Tarkovski directed Solaris in 1972, Stanley Kubrick directed a Space Odyssey in 1968, and Fred M. Wilcox directed Forbidden Planet in 1956.

Chapter 9: Facts About Rockets

"The rockets... can be built so powerfully that they could be capable of carrying a man aloft." **- Hermann Oberth**

The Word "Rocket" wasn't Coined Until the 17th Century When It was First Employed

The word "rocket" derives from the Italian rocchetto, which first appeared around 1610. On the other hand, the word rocket ship appeared in print in 1927.

Rockets Were Initially Employed in Combat for the First Time

The Chinese military came up with the idea for a gunpowder-powered arrow during the conflict between their empire and the Mongol invaders. These projectiles were reported as "arrows of flying fire" by the Mongols as they withdrew from the area. The first-ever rockets to use solid fuel were these fire arrows. Following these initial steps, more experimentation with cased gunpowder led to the invention of rockets and missiles for use in the military.

You Can Launch a Rocket from Your Backyard

Learning how to construct a rocket can help you better understand the science that goes into it. It's possible you won't believe this, but you can build a basic rocket from things you already have lying around the house. You can show the mechanism and principles of a rocket with nothing more than a piece of paper and a fizzy tablet.

The Quantity of Thrust is Used to Measure the Power of a Rocket

Rockets are projectiles that can push themselves forward. The exhaust from the rocket ensures that it continues along the same path even as it burns up the fuel stored in the body. The term "thrust" refers to the force exerted by a rocket's engine. Because rockets weigh millions of kilograms, they are necessary to generate the equivalent amount of power to break through and escape the earth's atmosphere.

In 1942, Germany Successfully Launched the First Rocket into Space

In 1942, Germany successfully launched the first rocket that could travel into space. Although it was known as the V-2 rocket, its primary function was not to carry people into space. In its place, a ballistic missile version of the V-2 was developed during World War II. Despite this, it was discovered during a test flight to be the first man-made object to successfully launch into space.

What Rockets are and How They Work in Practice

Early rocket technology was used in military engagements and fireworks displays across Asia and Europe by the 16th century.

According to NASA, one hundred years later, in the 17th century, a Polish-Lithuanian officer named Kazimierz Siemienowicz wrote his book titled "The Complete Art of Artillery." In this book, he contained concepts for rockets and multistage rockets.

In a concise reference to rocketry, the agency noted, "He published a design for multistage rockets that were to become core rocket technology for rockets traveling for outer space." Siemienowicz also suggested using batteries to launch military rockets and delta-wing stabilizers to replace the guiding rods already utilized with military rockets.

China Invented the First Rocket in the Year 1100 AD

Solid propellants were utilized in the rockets invented in China in the 10th century, and the rockets' primary functions were as weapons and pyrotechnics. The field of rocket science did not see a significant advancement in research until the 1920s. Rocket engineering as a professional discipline didn't get off the ground until the 1930s and 1940s.

The Height of the Giant Rocket Exceeds 100 Meters (300 Feet)

The Saturn 5 rocket developed by NASA is the largest one ever used. It has a height of 363 feet and a thrust of 7.6 million pounds. Between 1960 and the early 1970s, lunar missions were carried out using the Saturn 5 spacecraft.

In Chinese Mythology, There is a Chair That Can Be Propelled By Rockets

One of the first references to rockets as a mode of transportation may be found in traditional Chinese literature. Rocketry can be traced back to China. The story goes that a Chinese official named Wan-Hu built a rocket-powered flying chair at some point. The chair was outfitted with two enormous kites on either side, each of which had 47 arrow rockets attached to it.

You can probably figure out where I'm going with this. Wan Hu ordered his attendants to light the 47 fire arrows just before the scheduled takeoff time. The arrows were successful in their mission, but when the smoke cleared, neither Wan Hu nor his chair could be found anywhere. The joke is that even if it did take place, it is pretty likely that the explosion entirely obliterated him. Unquestionably one of the more amusing aspects of rocketry.

Rockets That Use Liquid Fuel are More Potent Than Rockets That Use Solid Fuel

China is credited as being the country that initially invented solid-fuel rockets in the 13th century. But in the end, it was rockets powered by liquid fuel that was responsible for opening the door to space flight. Rockets that use liquid fuel have more reactive oxidizers than those used in solid-fuel rockets. This results in a more potent chemical reaction, generating higher thrust. As a result, it takes a spacecraft traveling at 4.9 miles per second to escape the pull of Earth's gravity, which is close to the top speed achieved by a factory sports vehicle anywhere in the globe.

Solid-Fuel Rockets are Safer

When you have a lot of power, you also have a lot of duty and sometimes a lot of risks. Liquid-fuel rockets include more chemical components. Thus, it's more volatile. In addition, it is more expensive to safely store and manage liquid fuel rockets.

Rescue Efforts Have Sometimes Involved the Use of Rockets

Rockets have a variety of applications, including not only space flight and ammunition for the military but also search and rescue. For example, in the 19th and 20th centuries, rockets launched lines to inaccessible damaged ships. A buoy was passed through these lines and made available for passengers to grab to make their escape. Ships even used to have what was known as a rocket brigade, which consisted of a group of people who were responsible for firing rockets.

Rockets Launch in 3 Stages

Rockets are not composed of a single piece of solid metal. You may have noticed how a rocket in space breaks apart if you've ever seen it depicted in cartoons or on television shows. In essence, a rocket disassembles into its parts as it is in flight. Ignition is the initial step in the process of launching a rocket. After liftoff, the rocket will be propelled forward by the exhaust from the burning rocket. Once there is no more fuel, the part will separate from the rest. From that point on, the fuel reserve for the second stage will be utilized until the cargo enters orbit.

Rockets are Noisy

The roar produced by the Rocket is so intense that it has the potential to bring it down. Acoustic waves can produce up to 180 decibels of sound during the launch of a space rocket, which is almost as loud as an explosion caused by a nuclear weapon. The launch pad was constructed with a sound-dampening system to prevent this from happening. The system, comprised of water, soaks up the sound waves produced by the Rocket during launch.

Of All the Rockets Launched into Space, the Saturn V Rocket had the Highest Success Rate

In addition to being the giant Rocket, it also holds the record for the most successful Rocket, having completed 13 launches. At the moment, at least two stages of the Saturn V rocket used during the Apollo missions are still floating around in orbit. They are still considered Near-Earth Objects by NASA, which is tracking them. Fantastic information about rockets.

Rockets Have Traveled To the Location in Space that is the Farthest Known from Earth

As of 2018, Voyager 1 and 2 are located approximately 21 billion kilometers (13 billion miles) from Earth. The two rockets are now the human-made things farthest away from the planet. The satellites entered a region known as the heliosheath at the beginning of 2009, a part of space that had never been viewed. They are currently moving in the direction of the heliopause, which is the physical boundary between the solar system and interstellar space.

A Wooden Duck from Ancient Greece is Credited as Being the First Object To Demonstrate Rocket Technology

Even though it wasn't exactly a rocket, this experiment is widely regarded as the first to establish rocket science fundamentals. In addition, an individual named Archytas is said to have used a steam exhaust to guide a wooden pigeon through a line of obstacles in ancient Greek literature.

During the Launch of the Discovery Spacecraft, Elton John Played a Rendition of His Famous Song "Rocket Man"

At the Space Shuttle Discovery launch site in 1998, the iconic artist performed the classic song with the fittingly named title track.

In the Comics, There is a Character Named Rocket

The Beatles' song "Rocky Raccoon" would appear to be the source of inspiration for the raccoon that lives in Guardians of the Galaxy. What do you think of that as rocket science?

Huntsville in Alabama is called "Rocket City"

Rocket City is the nickname given to the city of Huntsville, Alabama, in the United States. This is because the city is considered the birthplace of the space program in the United States. Huntsville was responsible for the construction of all of the rockets that were used in the Lunar missions. In addition, Huntsville is the location of the present headquarters of NASA's Space Camp.

Some Flares Work Like Rockets

Rockets can also be used to send out signals of distress. Rocket systems are used in the vast majority of current flares. A trigger and launching system similar to that of spacecraft and missiles is utilized in rocket flares.

The First Person To Die Due To Space Travel was on a Rocket

The first person to die in an accident connected to space travel was a crew member of a Soviet rocket that burst at the Baikonur Space Center in Kazakhstan. One hundred sixty-five persons were killed in the blast, including the commander-in-chief of the Soviet rocket forces. The accident was not admitted to having occurred until 1990.

The Space Race was Responsible for Many Rockets Explosions

During the 1960s, there was a period known as the "Space Race," in which the United States of America and the Soviet Union participated in fierce competition. During an era when rocket engineering had not yet been perfected, numerous attempts at flight testing ended in catastrophic failures.

Rockets Generate a Great Deal of Heat

The temperature in a rocket's main engine combustion chamber can reach up to 3,315.6 degrees Celsius while it is operational.

There is No Room for Error When Dealing with Rockets

One of the key points to remember about rockets is that the catastrophic explosion of the Space Shuttle Challenger resulted in the deaths of seven crew members. After investigation, it was determined that malfunctioning O-rings in the right solid rocket booster of the shuttle were the cause of the unexpected explosion.

The Use of Rockets Assists in the Collection of Data

When deployed into space, sounding rockets carry equipment that may take data from a distance of 50 kilometers (31 miles) to 1,500 kilometers (930 miles) above the surface of the Earth. These rockets are primarily responsible for the majority of the knowledge that we have on space.

William Leitch was the First to Put Out the Idea of Utilizing Rockets for Space Travel

Leitch introduced the concept of space travel powered by rockets for the first time in his essay titled "A Journey Through Space," which was published in 1861.

A Rocket Ship can Overcome the Weightlessness of Space

There is neither air nor an atmosphere in space; it is empty. A spacecraft's engines allow it to travel through the vacuum of space. The propulsion in rocket engines comes from a chemical reaction, which causes the engine to move forward. The exhaust from the rockets used to power the ship is released in the opposite direction and at high speed. This makes it possible for the ship to find its way across space and maintain its trajectory. That is undeniably one of the truths about rockets you need to remember.

Rockets are Sensitive

One of the more unsettling truths about rockets is that even a very harmless mistake can have terrible consequences for both the Rocket and the crew aboard it. Only one instance of a Soyuz rocket making an emergency landing has been documented. To our great relief, none of the crew members were hurt. On the other hand, an inquiry found a malfunctioning sensor that had been destroyed during the assembling process.

Cargo Shipments to the International Space Station are Typically Carried Out By Rockets

Since there is no way for anything like FedEx to reach space, NASA is forced to rely on rockets to get supplies back to Earth. The proliferation of 3D printing technology, on the other hand, has made a cheaper and more effective alternative available.

SpaceX is Planning To Establish a Colony on Mars

Elon Musk is the mastermind behind Space Exploration Technologies Corp., a company whose mission is to locate and investigate the possibility of life on Mars as quickly as possible using rockets. However, SpaceX also provides cargo delivery to the International Space Station.

Chapter 10: Facts About Satellites

"I don't think there's anything on this planet that more trumpets life than the sunflower. For me, that's because of the reason behind its name. Not because it looks like the sun but because it follows the sun. During the course of the day, the head tracks the journey of the sun across the sky. A satellite dish for sunshine. Wherever light is, no matter how weak, these flowers will find it. And that's such an admirable thing. And such a lesson in life." - **Helen Mirren**

NASA Has Two Satellites Following Each Other Around the Earth, Measuring the Distance Between Them

NASA sent a pair of "Tom and Jerry" satellites into space in 2002 as part of the GRACE mission. These satellites had been made to last only five years, but they had been up for much longer. They sent back a lot of information that helped scientists figure out things like rising sea levels, melting ice sheets, and droughts. The batteries, sensors, and cameras on the new GRACE-FO probes are even more robust, so they can track much more accurately.

The GRACE-FO mission will use Earth's gravitational field to map where water is on the planet's surface. This will add a new dimension to remote sensing in space. Water on the ground is a great way to see how the land changes, whether in the form of melting ice (which makes glaciers smaller) or heavy rainfall (which pumps up underground water reservoirs). Changes in water levels cause tiny changes in land mass, which affect the gravitational field, which moves ever so slightly and can be measured by satellites regularly. This will help scientists make an accurate map of the world's water. For example, we know that about a third of the world's groundwater is being used up because of the GRACE satellites that came before. And that the rate at which land ice melts in places like Greenland and Antarctica is getting much faster every day.

Strangely, the pull of Earth's gravity on a satellite in orbit is weaker where the elevation is lower and stronger where the elevation is higher. So, gravity is more robust in places like the south Pacific Ocean, and the mantle under the Earth's crust is thicker.

Sarah Parcak and Her Team Found 17 Pyramids, 1,000 Tombs, and Over 3,100 Towns in Egypt Using Trial-and-Error Algorithms on High-Resolution Satellite Photos

Sarah Parcak is known as "the Indiana Jones of the 21st century" because of her innovative work in space archaeology. When she went on The Late Show with Stephen Colbert in January and wore Indian Jones's hat, the host called her a "space archaeologist wizard genius." She won the 2016 TED Prize and $1 million in November 2015. Parcak is a pioneer in her field. She uses old military satellites to find ancient ruins that had not been seen before. Parcak and her team have found 17 pyramids, 1,000 tombs, and more than 3,100 settlements in Egypt alone by looking at high-resolution satellite images with algorithms they made by trying things out and seeing what works and what doesn't. Now, she wants to use the tools of space archaeology to help keep these sites safe and ensure that humanity's cultural history isn't lost.

Satellites are Set Up To Stay Away from Meteorites

Only one of the roughly 8000 satellites that have been launched has been hit by a meteor and destroyed by it. This is surprising because there would be a lot of opportunities for this to happen. In 1993, this happened to the communications satellite Olympus, owned by the European Space Agency. Satellites are set up to avoid things like the Perseid meteor showers, but the Perseids moved that year.

It Takes Pretty Advanced Technology To Hit a Satellite from the Ground

It was 500 miles above the surface of the Earth. The explosion sent a debris cloud into space, adding to the "space junk" that circles the Earth. Hans Kristensen, a weapons expert at the Federation of American Scientists, says that even though it was thought China was working to make these kinds of weapons, he was still surprised by the satellite attack. U.S. officials say the Chinese used a medium-range ballistic missile, most likely the DF-21, to hit the satellite. A rocket can hit a satellite, which is a pretty minor thing. Kristensen estimates the DF-21's accuracy at a few hundred feet. The satellite was probably smaller than a fridge.

The Hitomi X-Ray Telescope

The Hitomi X-ray telescope blew up in space because of a bug in the software that made it try to make up for a rotation it didn't have. This overcompensation started a chain reaction that caused the satellite to spin out of control and blow up.

The First Time the US Tried To Send a Satellite into Space, It Didn't Work

The first time the United States tried to send a satellite into space, on December 6, 1957, it failed and exploded. The Naval Research Laboratory's Vanguard TV3 (Test Vehicle 3) was a small satellite made to test the three-stage Vanguard rocket's ability to launch and to study how the environment affects a satellite and its systems while it is in Earth orbit. It was also meant to be used for orbit analysis to get geodetic measurements. At launch, the rocket went up about 4 feet, but then it lost power and fell back to the launch pad at Cape Canaveral in Florida. The fuel tanks then broke and exploded, destroying the rocket, damaging the launch pad, and making the 3-pound satellite useless.

The US could not have picked a worse time for the launch to fail. It was in the middle of the Cold War with the Soviet Union, and Sputnik 1 had just beaten it to space. The US put out a lot of information about the launch, making its failure more evident to other countries and its people.

The event was called "Flopnik" and "Kaputnik" in the news. Because of the failed launch, the New York Stock Exchange stopped trading in the stock of the Martin Company, which was the main contractor for the project. In the end, the scientific goals of the Vanguard Project were met with the success of the March 17, 1958, launch of Vanguard 1. The mission taught us a lot about the Earth's size and shape, the air's density, the range of temperatures, and how micrometeorites hit the Earth. In 1958, NASA took over responsibility for Project Vanguard. With the launch of Vanguard 3 in 1959, the program ended.

Sputnik 1 was Launched on October 4, 1957

On October 4, 1957, the Soviet Union's Sputnik 1 was launched from Baikonur, Kazakhstan, into an elliptical low-Earth orbit. It was the first artificial satellite to work and started the Space Race. The satellite moved at a speed of about 29,000 kilometers per hour, and each orbit took 96.2 minutes. It sent signals on 20.005 MHz and 40.002 MHz, which amateur radio operators worldwide could listen to. The Soviet Union and several amateur radio groups pushed for these kinds of monitoring efforts. The satellite was a sphere with a diameter of 585 mm. It comprised two hemispheres sealed with O-rings and connected with 36 bolts. It was 83.6 kilograms in weight. The satellite had two antennas, each with a nearly spherical radiation pattern. This meant that the satellite beeps were sent out with the same amount of power in all directions, so the signal could be received even if the satellite was moving. There were three silver-zinc batteries in the power supply. Two batteries were used to power the radio transmitter, and one was used to power the system that controlled the temperature. They were supposed to stop working in two weeks, but they worked for 22 days, so the signals could continue until October 26, 1957.

Unfortunately, Sputnik 1 burned up on January 4, 1958, as it fell out of orbit and back into Earth's atmosphere. It had been in rotation for about three months and traveled about 60 million km (37 million miles).

People in the United States were shocked by the success. Because of propaganda, many Americans thought the US was technologically better than its Cold War rival. Even though many other things were going on, many people thought that with Sputnik 1, the Soviet Union had reached new and unique technological heights, giving it an edge in exploration, military, and defense.

Satellite Color

The kind of light a satellite instrument measures determines the colors in an image. True-color images use light with red, green, and blue wavelengths, so the colors look like what you would see from space. False-color images use infrared light, making colors look different than expected.

Earth Has Satellites That Go Around It

There are thousands of satellites going around Earth in the sky above us. Satellites can help the government, the military, and even regular people in many ways. They make it possible for us to use things like the Internet, TV, GPS, and much more. They also have scientific uses, like looking at the Earth and space, making it possible for high-level technology to be made. More than half of the 4,550 satellites that orbit Earth are used for communication, and that number will keep rising as tech billionaires try to bring high-speed Internet to every part of the planet.

In 1979, the Skylab Satellite Crashed in Australia

In 1979, the Skylab satellite crashed in Australia, sending debris into the Indian Ocean. This happened five years after the last person worked on Skylab. Not one person was hurt. Skylab was the first space station to work well. It was launched into space in 1973. The first manned mission to Skylab happened two years after Salyut 1, the world's first space station, was put into orbit around the earth by the Soviet Union. But unlike Salyut, which always had problems, the American space station was hugely successful. It could keep three separate crews of three people safe for long periods.

Satellite vs. Spacecraft

A satellite is an object that goes around the Earth, the moon, or some other celestial body. A spacecraft is a vehicle or device made to travel or work outside the Earth's atmosphere. The Chandra X-ray Observatory and the Hubble Space Telescope are artificial satellites because they are spacecraft in orbit around the Earth. The moon is one of Earth's natural satellites, just like Jupiter's moons are one of Jupiter's natural satellites. Voyager and Pioneer are not satellites because they are not in orbit around any other object.

The International Space Station (ISS) is the Man-Made Object with the Most Mass

A satellite made by people and put into orbit by rockets is called an artificial satellite. More than 3,000 working satellites are in orbit around the Earth right now. The purpose of a satellite determines how big it is, how high it goes, and how it is made. The International Space Station is the man-made object with the most mass (ISS). The central part of this is as big as a five-bedroom house, but when you add in the solar panels, it's as big as a rugby field.

Only One Natural Satellite is in Orbit Around the Earth, and That is the Moon

Only one natural satellite is in orbit around the Earth, the Moon. It is the largest and most massive satellite compared to its parent planet, making it the fifth-largest satellite in the Solar System overall. However, it is only the fifth largest satellite overall in the Solar System.

Conclusion

Incredible things can be found in space. However, the sheer magnitude of it, together with the trillions upon trillions of objects spread across the many billion galaxies that make up the universe, is nearly incomprehensible to the brain of a human being.

This book has exposed you to the amazing things you can find in space. I hope you learned a lot!

Made in United States
Troutdale, OR
11/03/2026

56852695R00082